Westem Food Classroom II

西餐教室II

Western Food Classroom II

西餐教室II

辽宁西餐专业委员会◎编

吉林科学技术出版社

图书在版编目（C I P）数据

西餐教室Ⅱ / 辽宁西餐专业委员会编. -- 长春 : 吉林科学技术出版社, 2017.7
ISBN 978-7-5578-1789-3

Ⅰ. ①项… Ⅱ. ①辽… Ⅲ. ①西餐－食谱 Ⅳ. ①TS972.188

中国版本图书馆CIP数据核字(2017)第007732号

西餐教室Ⅱ XICAN JIAOSHI II

编　辽宁西餐专业委员会
出 版 人　李　梁
策划责任编辑　张伟泽
执行责任编辑　王运哲
封面设计　长春创意广告图文制作有限责任公司
制　　版　长春创意广告图文制作有限责任公司
开　　本　889 mm×1194 mm　1/16
字　　数　480千字
印　　张　15
印　　数　1-3 000册
版　　次　2018年1月第1版
印　　次　2018年1月第1次印刷
出　　版　吉林科学技术出版社
发　　行　吉林科学技术出版社
地　　址　长春市人民大街4646号
邮　　编　130021
发行部电话/传真　0431-85635177　85651759
85651628　85635176
储运部电话　0431-86059116
编辑部电话　0431-85670016
网　　址　www.jlstp.net
印　　刷　吉林省创美堂印刷有限公司
书　　号　ISBN 978-7-5578-1789-3
定　　价　79.90元
如有印装质量问题可寄出版社调换
版权所有　翻印必究　　举报电话: 0431-85635186

编 委 会 Editorial Board

专家顾问 Expert Advisor

张奔腾

主　　编 Chief Editor

亢　亮　丁建军　马　翀

副 主 编 Associate Editor

安　阳　李　季　彭　刚　刘　鹤　孙　强　张　平　马海波　姚国庆　田　达
李　杨　郭东海　张岩松　蔡　巍　黄学志　王伟诗　韩　笑　文　浩　韩　冬

编　　委 The Editorial Board

司连福　曹　阳　党宏坤　吴　钢　钱　丰　陈　颂　孟　宇　高大伟　吕　放
李晓峰　刘　冬　陈　旭　关　威　刘　贺　陈　辉　种　磊　王德义　张德斌
马洪彬　张　旭　赵　强　魏上桀　刘宗喜　邓昌丰　柴彦丰　车德蒙　杨文飞
杨　宇　王松鹤　靳蔚君　吴　艳　张宇涛　王　刚　贾子明　张　楠　艾圣凯
郑超远　冯玉杰　赵华阳　宋　萌　郭　清　关　冲　付春贺　刘　强　蔡云龙
马　超　孙　鹏　孟宇飞　冯　辉　李东生　刘　军　李　响　刘　博　段　辉
郭　航　蒙丹阳　藤典军　彭　程　余欣佳　贾小东　于　月　霍　宁　张　羽
曾凡东　王丽娜　方　圆　王辰龙　史大旺　张王海　李　亮　袁晓楠　徐　跃
芦光宇　谭龙巍　阮　详　闫　寒　赵　亮　刘　双

英文翻译 English Translater

董建明

序 言

在追求个性化、时尚化的当今社会，西餐曾经被视为小众的产品正逐渐走向主流，并已经成为我国高星级饭店和社会餐饮的重要组成部分。

为了满足辽沈地区饮食文化和西餐餐饮创新发展的需要，辽宁省饭店餐饮协会于2015年初创立了辽宁西餐专业委员会，并把编撰一部西餐烹饪菜典作为专业委员会的要务之一。经过专业委员会全体成员及西餐大师、餐厅总厨、厨师长等业内精英近一年的辛勤耕耘，《西餐教室》终于面世了。本书坚持实用性和专业性相结合的原则，着重介绍了辽沈地区高星级饭店和社会餐饮正在 主打的经典热销西餐菜式。其中既保留了正统的西餐特色，又迎合了现代餐饮的时尚潮流。让我们在领略辽沈西餐文化的同时，也能更好地品味辽沈美食。

本书收录了98位西餐工作者的多年烹饪心得，其中展示的西餐菜肴制作工艺精湛，菜系流派纷呈、各具风味，是一部铸就辽沈西餐烹饪文化的经典之作。随着经济全球化及互联网+时代的来临，中西方饮食文化将会相互交融，取长补短，人们将会享受到更美味、更快捷、更营养的西餐食品。同时我们希望在当前大众创业、万众创新的时代背景下，本书也能对今后餐饮创新发展有所帮助。

《西餐教室》能在较短的时间内顺利出版，得益于吉林科学技术出版社的大力支持，在此一并表示感谢！

辽宁省饭店餐饮协会
张宝学

Preface

In such a society pursuing the individualism and fashion, western food once considered niche products are gradually towards the mainstream, and has become an important part of star hotels and social dieting.

In order to meet the development of food industry and western food innovation in Liao Ning Province, Liaoning Restaurant And Catering Association established Liaoning Western-style Food Professional Committee in 2015 and set the goal to publish a book about western food cooking as its priority. Western Food Classroom finally is released. This book combines the goal of being both practical and professional, focusing on introducing the popular and classic western dishes in the star-hotels and social dieting in Liao Ning Province, which keeps some of the traditional characteristic of western dishes and at the same time meet the fashionable tendency of modern dieting development. It will enable us not only to appreciate the western food culture but also taste the delicious food in Liao Ning Province.

With the many years' working experience of 98 western food chefs, this book is going to present us one of the important characteristic of western dishes that appears to be delicate processing, a variety of dishes and taste. It could be regarded as the classic work establishing the culture of western food cooking in Liao Ning Province. With the coming of economic globalization and Internet+, Western and Chinese dieting will mix and take advantage with each other. People will enjoy the more delicious, convenient and nutritious western food. Meanwhile we also hope that this book will contribute to the development of food industry in the era of venturing and innovating.

Acknowledge the support of Jilin Science and Technology Publishing House to promote the release of Western Food Classroom in such short time.

Liaoning Restaurant And Catering Association

Zhang Baoxue

目 录 Contents

第一章：西餐门徒 Cooking Apprentice

第二章：西厨的秘密 Secrets in the Kitchen

第三章：头盘类 Appetizers

第四章：汤类 Soup

第五章：主食类 Staple Food

第六章：肉类 Meat

第七章：日餐类 Japanese Cooking

第八章：甜品类 Dessert

第一章：西餐门徒 Cooking Apprentice

第一节：厨房分工 Division of Labor

西式厨房的分工相当仔细，现介绍一个标准的西厨分工结构，让年轻厨师有个初步认知概念。

The division of labor in the Western-style food kitchen is quite meticulous. Now the standard division of kitchen labor will be introduced so that the young cooks could have a preliminary concept.

1. 西厨职位分工结构表 Classic Kitchen Organization Chart

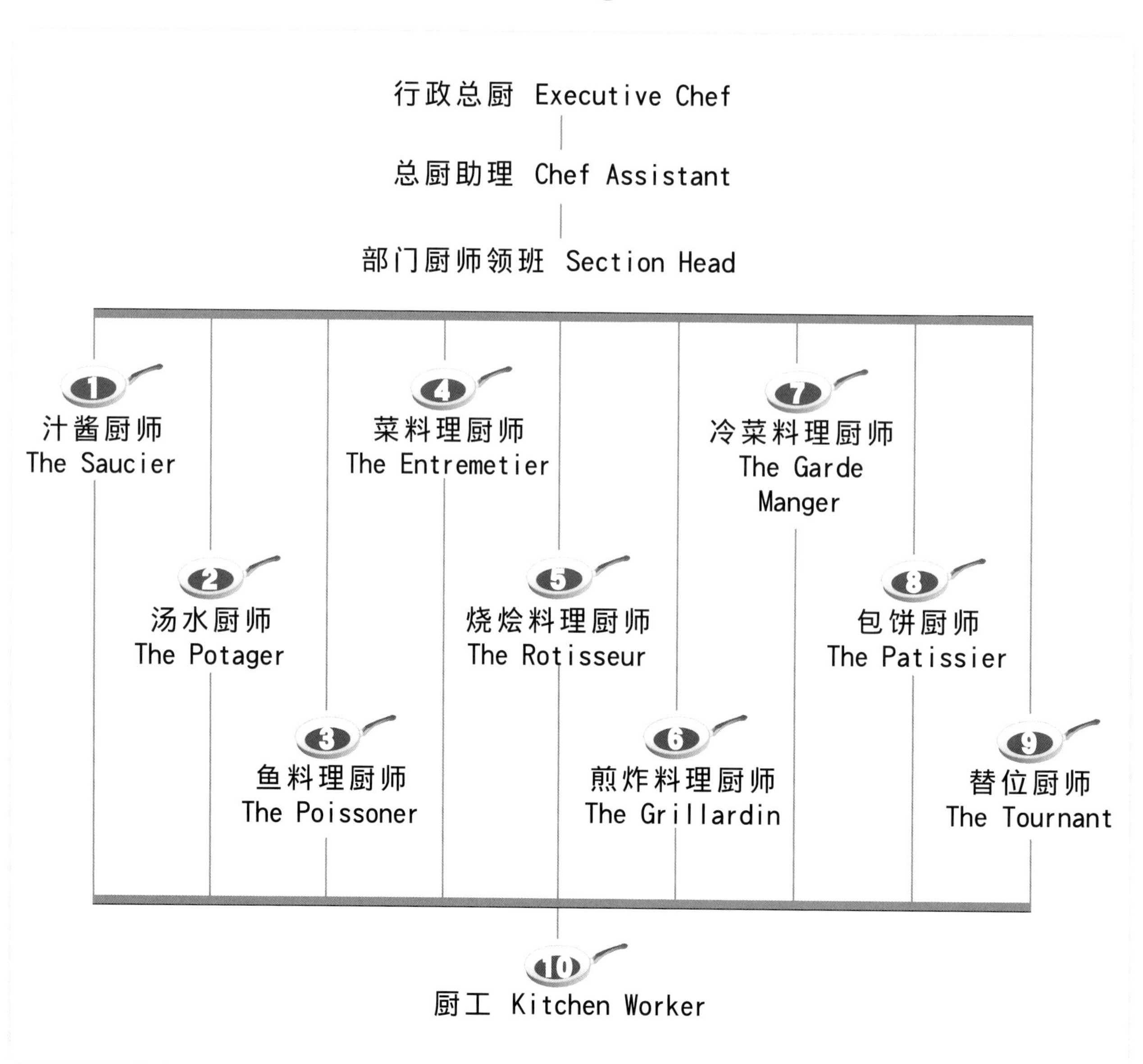

2. 厨房员工的权责 Job Description

行政总厨：是厨房的灵魂人物，负责所有对外及对内的工作，监控整个厨房的运作，设计餐单、控制成本与餐厅经理紧密联系等。

Executive Chef：As the soul of a kitchen, the executive chef is responsible for managing internal and external affairs, monitoring the operation of the whole kitchen, designing menus, controlling the cost and keeping close contact with restaurant's managers.

总厨助理：负责协助总厨日常工作，监控所有生产过程、出品质量及员工培训等。

Sous Chef：Responsible for helping the executive chef in daily operation, monitoring every production process and quality of products as well as training of staff.

部门厨师领班：负责各部门的主要日常运作，特别是生产管理。

Division Foreman：Responsible for the routine operation, especially the production.

汁酱厨师：主要负责制作汁酱、汤底、热头盘及炒类料理。

The Saucier：Reponsible for making sauces, stocks, hot appetizer and sauté dishes.

汤水厨师：主要负责汤底及汤类的料理工作。

The Potager：Responsible for making stock and producing dishes from broth.

鱼料理厨师：主要负责鱼类及海鲜类的料理。

The Poissoner：Responsible for producing dishes of fish and seafood.

菜料理厨师：负责蔬菜、淀粉质类如薯仔及蛋类的料理制作。

The Entremetier：Responsible for producing dishes of vegetables, starch (like potatoes) and egg.

烧烩料理厨师：负责烧制肉类、先烧后烩的料理及肉汁制作。

The Rotisseur：Responsible for making roast meat, dishes which are first roasted and then braised as well as meat sauce.

煎炸料理厨师：负责所有煎或炸的肉类及海鲜料理制作。

The Grillardin：Responsible for pan-frying and deep-frying meat and seafood.

冷菜料理厨师：主要负责所有冻肉及菜的料理，包括沙拉、沙拉汁、冻酱糕、冻头盘及自助餐的头盘料理制作。

The Garde Manger：Responsible for all cold meat, vegetables dishes, including making salad, salad sauce, cold sauce, cold appetizer and appetizer for buffet.

包饼厨师：负责所有甜点及面包类的制作。

The Patissier：Responsible for making desserts and bread.

替位厨师：负责当其他厨师休息时代替其位置。

The Tournant：When the chef is on leave, the tournant will replace them.

厨工：初级厨师主要负责协助各部门的厨师制作料理及搬运货品、清洁厨房及用具等工作。

Commis Cook：A junior cook who is responsible for helping the chef of each department to prepare dishes, transport goods, clean the kitchen and tools.

第二节：基本烹调方法 Basic Cooking Skills

炒：炒时要用猛火，因为肉的蛋白质遇到火会凝固，用猛火可使肉的表层收缩，封住肉内的汁液不让它流失。

Sauté：Use high heat during sautéing. Protein in meat solidifies over high heat, when meat is cooked this way, the skin layer of meat contracts and hence meat juice is sealed and will not flow out easily.

煎：食物以鑊及少量热油来烹调，食物表面直接与热油接触。需要保留肉中汁液才能做出美味的菜式，可于调味后加上薄层面粉来保护或以猛火煎封肉面均可。

Pan Fry：During pan frying, food is cooked with little hot oil in a wok or pan, surface of food touches hot oil directly. To make a luscious dish, coat meat with a thin layer of flour or fry meat over high heat so as to keep the juice.

炸：食物全部接触热油来烹调。此方法会使部分油脂在烹煮过程中渗入肉中，较适宜脂肪含量较低的材料；另外要留意的是油温，油温太低会引致过多油脂渗入食物之中，影响食味及效果；可蘸上面包糠或粉浆作为保护层，同时亦能将肉汁保留。

Deep fry：during deep frying, the whole piece of food is soaked in hot oil. In this way some oil penetrates into the meat, so ingredients containing low fat are more suitable for deep frying. In addition, be aware of oil temperature. Taste of food will be affected if it is deep fried in low-temperature oil as much oil will penetrate into the piece of meat. In this case, coat meat with thin layer of breadcrumb or flour to keep the juice.

烧烤/扒：此烹调法适用于较高质素及较薄的肉类，如西冷牛排；将肉食置于炉架上以炉火直接烹煮，如烹煮含较少脂肪的肉块如牛柳时，便应扫上少量油脂以避免其干焦，此方法能在较短的时间内煮熟食物，同时又能保存肉质本身的味道。海鲜亦能以此法烹调。

Grill：It is good to grill high-quality or thin piece of meat like sirloin steak. Put meat on grid and cook over fire directly. Brush some oil meat with less fat, like beef tenderloin, to prevent scorching. Grilling can cook meat within a short period of time and help to preserve its original taste. Seafood can also be grilled.

烘烤：烤是指将食物放置于焗炉内烹调，食物是以空气对流的方式传导热力。食物中一定的水分会流失，所以不适用于太小的肉类，所烧的肉类应最少为500克以上；开始时应用猛火，再转为较低的火力去完成所需的热度。

Roast：It means putting the food in an oven, food is cooked by heat circulation. During roasting, water in food is lost. Meat lighter than 500 g is not suitable for roasting. Start roasting with high heat, then switch to low heat for a better result.

烩：是指以适量液体，用较长时间来烹煮食物，要诀是以强火煮开后再以慢火来烩。这种方法能有效地保留食物的养分及香味。需要注意的是，应用较韧的肉来做，因肉味较浓而烩后肉件不会散开；同时应加上煲盖烹调，以保留食物的水分及香味。

Stew：Stew is to cook food in liquid over a long period of time. Start to boil liquid with high heat then stew over low heat. By stewing, nutritions and fragrance of food can be kept. Use tough meat as its flavour is stronger and will not fall apart after stewing. To keep the moisture and aroma, cover with a lid during stewing.

蒸：以蒸气来烹煮食物能充分地保存食物的养分。特别适合注重健康的人士，但缺点是需要较长的烹调时间。

Steam：Nutrition of food can almost be kept when steamed. Steamed food is light in flavor, tender and smooth and so it is good for people who prefer a healthy diet. Long cooking is however a disadvantage.

浸煮：是指将食物置于微滚的液体中烹煮，可使用水、汤或酒等；适用于肉质较嫩的食物，如海鲜或水果等。这方法能有效地保存食物的香味和嫩滑度。

Poach：Poach means to cook food in slightly boiled liquid like water, soup, wine, etc. Poaching is suitable for food with soft texture like seafood or fruits, and can effectively preserve the fragrance and the tender texture of meat.

第二章：西厨的秘密 Secrets in the Kitchen

第一节：常见工具 Common Tools

常见工具（量杯、勺等）：手是最完美的工具

Common tools (Measuring Cup, spoon, etc.): Hands are the perfect tool.

量匙：量度微少食材的量器，主要分为茶匙和汤匙。清晰刻度明确标示容量，易识别，准确计算食物份量。

Measuring Spoon: Small amount of dry ingredients are measured in teaspoons and tablespoons. The scales for the capacity are clear and easily recognized. The amount of the food can be correctly calculated.

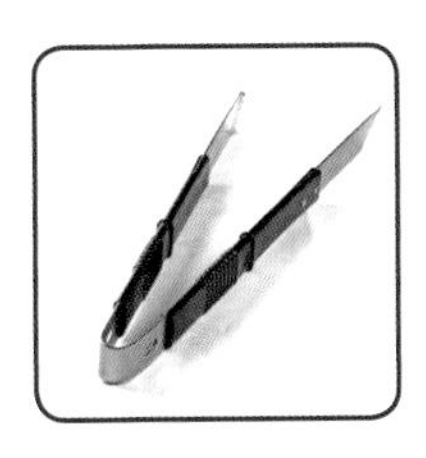

食物钳：以夹取食物或作烹调食物的器具，形如V字，属不锈钢原料。

Tong: They are used to pick up and transfer food. Its shape is just like a "V" and it is made of stainless steel.

量杯：量度液体的量器，每量杯约为250毫升（英式）。

Measuring Cup: It is commonly used to measure liquid. In general, one measuring cup is 250 ml (UK style) .

磨刀棒：磨刀棒不是刀，但却是刀具中不可缺少的一部分，用来磨刀，保持刀刃锋利。

Knife Sharpener: it is not a knife but is an indispensible part in the chef' s knives, which is used to sharpen the knives and keep is sharp.

厨刀：厨刀是厨房中最常用的刀具，用来切块、片、丁等。靠近刀柄部位宽，渐渐变窄，前端是尖形的。刀片长260 mm，最适宜日常使用，稍大的适宜于切片、块，小的适宜于做细中工。

Chef' s knife: As the most common tool in the kitchen, it can be used to cut, slice and dicing, etc. It holds the shape of being wide at the handle, then becoming narrow gradually with a pointed shape at the front end. With the 260mm blade, it is most suitable for the daily use. The larger are for slicing and dicing, and the smaller for dilicate work.

手动打蛋器：一般在比较容易搅拌的情况下使用，如搅拌西餐沙拉中的油醋汁，以及搅拌鸡蛋、黄油和一些简单的面糊等。手动打蛋器只靠手的力量来搅拌，所以很难把蛋清或鲜奶油打发。根据烹调需要，还可选择自动打蛋器打发材料。

Manual egg beater: It is used to stir in the easier situations such as stirring oil and vinegar of the salad in the western food or stirring eggs, butter and some simple batter. Since it is powered by hands, developing the egg white or the fresh cream could be very difficult. Automatic beater could also be chosen when necessary.

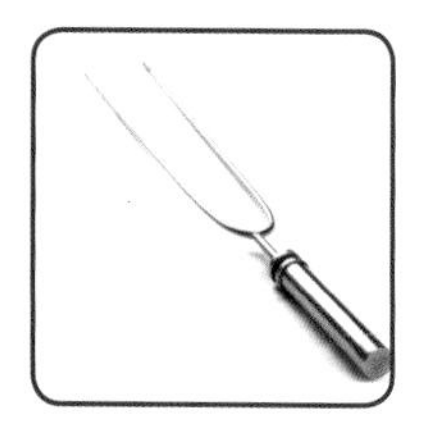

厨叉：它是长柄、两齿的叉子，分量重，用来拣起和翻转肉或其他食物，其做工一定要坚实，能叉起重物。

Chef's Fork: It has a long handle and two prongs with varied weight, mainly used to pick up or turn over meat or other food. In this way it must be made firmly in order to pick up heavey stuff.

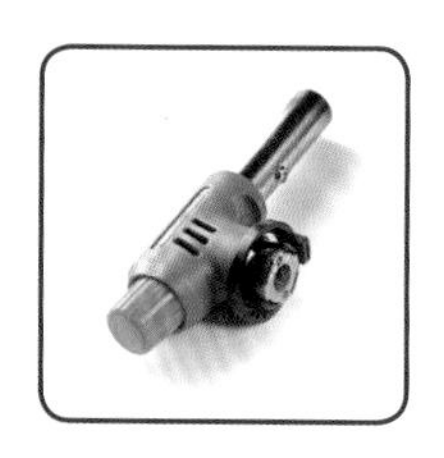

喷枪：主要用途是结合其他奶油食品造型，进行色彩处理的重要工具。

Spray lance: It is mainly used to shape food combined with other cream materials and deal with the color.

撇沫器：撇沫器长柄，微呈杯形。可过滤食物，勺片薄，用来撇去液体食物中的泡沫和固体碎屑。

Skimmer: it is with a long handle and a circle-shaped head, which could be used to filter food, skim the foam and solid fragments.

勺：大号不锈钢的勺，用来搅拌和使用；漏勺用来从液体中捞起固体食物。

Ladle: The large-sized ladle could be used to stir and be on table with the capacity. Colander could be used to pick up the solid from the liquid.

车轮刀：带柄、刀片可旋转的圆形刀具，主要用来切面团和烤熟的比萨。

Wheel cutter: It is equipped with a handle and its blade could be rotated mainly to cut dough and the roasted pizza.

温度计：用来测量烹饪食物的温度，便于随时检测原料的温度是否符合标准。

Thermometer: It is used to mea-sure the tempreture of the cooked food and see wheather the tempreture of the ingredients meets the standard.

滚针打孔器：做饼干或比萨时用来给面片扎洞，比较均匀，而且方便快捷。

Needle Punch: It is use to punch the facets when making biscuits and pizza. It could be used evenly and conveniently.

水果挖球器：用来挖出可爱的水果球，装饰在布丁、果汁里面。另一端可挖出原料的坏肉，亦可削皮。

Fruit ball digger: It is used to dig lovely fruit balls to decorate pudding or juice insides. The other end could be used to dig out the rotten meat or to peel.

第二节：常见食材 Common Ingredients

蔬菜	Vegetable			谷物与面	Cereal and Flour
大蒜	Garlic	甜椒	Bell Peper	意大利粉	Spagetti
红栎叶生菜	Red Oak Leaf Lettuce	蘑菇	Button Mushroom	直通粉	Penne
红珊瑚生菜	Red Coral Lettuce	冬菇	Shiitake Mushroom	螺丝粉	Gemelli Pasta
牛油生菜	Butter Lettuce	辣椒	Chilli	宽面条	Fettuccine
刁草	Dill	西蓝花	Broccoli	细面条	Vermicelli
卷叶苦苣	Curly Endive	茄子	Eggplant	野米	Wide Rice
罗勒	Basil	意大利青瓜子	Zucchini	印度米	Basmati Rice
百里香	Thyme	西红柿	Tomato	粗麦	Rough Wheat
青瓜	Cucumber	芹菜	Celery	意大利米	Risotto
				香米	Non-basmati

第三节：香草与香料 The Application of Herbs & Spices

1. 香料：能广泛地应用于各种料理之中，无论汤汁、肉类、海鲜，甚至甜点均能应用得到。

Herbs &Spices: they could be used widely in a variety of dishes , including soup, meat, seafood and even desert.

2. 香叶：能有效地去除异味，如鸡汤之中的腥味等；切忌过多，因其香味十分强烈。

Bay Leaf: It can remove odor like offensive smell of chicken soup. Don't add too much as it has a very strong smell.

3. 百里香：又名百搭香草，可用于各种料理之中，如汤汁、烩菜，并经常用于腌肉及海鲜料理中。

Thyme: It can be used in many foods like sauce and hodgepodge, always used in marinated meat and seafood.

4. 迷迭香：羊肉料理的最佳配搭。因味道强烈，所以只需较少份量即可；可加入汤汁或烩菜中，亦可使用于腌肉、腌馅、烧猪或烧羊腿之中。

Rosemary: It is a good partner of mutton. As it is strong in smell, just a small amount is to be used. It can be added in sauce ,hodgehodge or marinated meat, filling, roasted pig or roasted mutton.

5. 鼠尾草（洋苏菜）：常用于酿馅、牛仔肉或猪肉菜式之中，尤其适合于鸡肉料理。

Sage: It is usually used in filling, veal or pork, especially suitable for chicken.

6. 意大利香草：顾名思义，常用于意大利料理之中，特别适用于烩炖的菜式。

Oregano: It is always used in Italian dished, especially suitable for simmered dishes.

7. 胡椒：有白、黑、粉红、青等不同颜色，各有不同的味道，在料理中加入适量胡椒有画龙点睛之效。但在大多数料理中，仅用白胡椒和黑胡椒已足够。

基本上拥有这些香草与香料已足够派上用场了。用于烹饪上，可以去腥，增添菜肴香味。谨记香草与香料的正确使用方式是少放即可，因为其本义是要突出菜肴的香味而不是掩盖主菜的味道，这点必须牢记。

Pepper: There are white, black, pink and green peppers. Different clors of peppers contain different tastes. Just some pepper can make dishes brilliant. Usually, white and black pepper powder is already enough for seasoning.

Basically, having these herbs and spices is enough in the kichen, which could get rid of fishy smell and increase the taste.Remember, don't use too much herbs and spices, a small amount is enough as they are used to give prominence to dishes' fragrance , they are not used to cover the taste of main dishes.

第四节：葡萄酒的烹调应用 Cooking with Wine

在很多菜式的烹调中，都需要加入不同种类的酒，以使其香味融于菜式之中。

西餐中使用最多的是白葡萄酒，它是做鱼、海鲜、牛仔肉及鸡菜式时不可缺的。除了能有效去除海鲜的腥味，还能带出其原有的鲜味；做西餐时使用较便宜的白葡萄酒便可，因用几千元和几百元的白葡萄酒做了同样的菜式，得出的结果并不易辨别出。选购时应挑选干性的白葡萄酒，太甜的白葡萄酒会影响菜式的原味。

馥郁的红酒是牛肉的绝佳配搭，用作浸渍或烹煮均十分理想；红酒多用于肉类料理，也适用于一些特殊的海鲜烹调。

Many different kinds of wines are used during cooking so that fragrance of wine can be penetrated into foods.

White wine is mostly used in making western dishes. It is essential in cooking fish, seafood, veal and chicken.It cannot only remove unpleasant odor of seafood, but also release the freshmess and original taste. It is good enough to use inexpensive white wine to cook as cuisine cooked with expensive wine and relatively cheap wine cannot be distinguished easily. Choose dry white wine, if the white wine is too sweet, original tasted of dishes will be destroyed.

Perfumed red wine is a perfect partner of beef, it is ideal for both soaking and boiling. Red wine is usually used in cooking red meat , but it can also be used to cook some special seafood.

葡萄酒的用法一般分为以下几种：

1. 将酒煮熟后加入汤汁之中，或直接把酒加入汤汁之中。
2. 用作浸渍食肉。
3. 将已煎或烤好的海鲜或肉类淋上葡萄酒。
4. 用于正在煮制的菜肴之中。

Grape wines are generally used in the following cases:

1. Bring wine to the boil, add in sauce or add directly into sauce.
2. Soak meat in wine for use.
3. Sprinkle wine over fried or roasted seafood or meat.
4. Add in wine during cooking.

第五节：常用计量单位换算表 Convention Table of Measuring Units

液体 Fluid

1茶匙=5毫升	1 tea spoon(tsp)=5 ml
1汤匙=15毫升	1 table spoon(tbsp)=15 ml

固体 Solid

1茶匙=5克	1 tea spoon(tsp)=5 g
1汤匙=15克	1 table spoon(tbsp)=15 g

液体和固体 Fluid and Solid

1千克=1000毫升	1 kg=1000 ml
1升水=1千克	1 LWater=1 kg

温度表 Temperature Chart

小火	Low Heat	150 ℃
中火	Moderate Heat	180 ℃
大火	High Heat	200 ℃

白汤底 White Stock

材料 鸡骨750克，洋葱、胡萝卜、西芹（粗切）各50克，丁香、蒜头各2粒，白胡椒粒10粒，香草束1束，清水1.5升。

制作方法

1 所有材料放入锅内。

2 小火烹煮2.5小时，不时撇去汤上浮沫。

3 滤出清汤，用勺子压出汤渣内的剩余汤汁，待凉即可。

Ingredients

Chicken bone	750 g
Onion, roughly chopped	50 g
Carrot, roughly chopped	50 g
Celery, roughly chopped	50 g
Cloves, chopped	2 pcs
Garlic, chopped	2 pcs
White peppercorn	10 pcs
Bouquet garni	1 pc
Water	1.5 L

Procedures

1 Place all ingredients in pan and bring to the boil.

2 Simer for 2.5 hours and skim the foam frequently.

3 Strain the stock, press the solid with ladle to extract extra liquid and let cool.

布朗沙司 Brown Sauce

材料 牛骨1.5千克，番茄膏30毫升，洋葱100克，胡萝卜（粗切）150克，大蒜（粗切）80克，西芹（粗切）100克，胡椒粒8粒。

制作方法

1. 牛骨、洋葱、胡萝卜、大蒜、西芹置于烤盘上，放入220℃的烤箱内烤上色。
2. 将烤好的牛骨倒入锅中。
3. 加入番茄膏略炒。
4. 加入适量面粉拌匀。
5. 倒入3升清水煮制，一般需煮制8小时以上。
6. 滤出清汤，再从汤渣压出水分，待凉即可。

Ingredients

Beef bone	1.5 kg
Tomato paste	30 ml
Onion	100 g
Carrot, roughly chopped	150 g
Garlic, roughly chopped	80 g
Celery, roughly chopped	100 g
Peppercorn	8 pcs

Procedures

1. Place beef bone, onion, carrot, garlic and celery on the baked tray at 220℃ until it turns colored.
2. Put the baked beef bone into the pan.
3. Sauté slightly with the tomato paste.
4. Stir evenly with suitable flour.
5. Steam with 3L water for more than 8 hours.
6. Strain the stock, press the solid with ladle to extract extra liquid and let cool.

鱼汤底 Fish Stock

材料 鱼骨、洋葱、胡萝卜（均粗切）各100克，西芹（粗切）80克，干白酒200毫升，白胡椒粒10粒，月桂叶2片，柠檬1/2个，清水3升。

制作方法

1 鱼骨连同其他材料一同放入煲内煮滚。
2 转小火熬煮20分钟，撇去汤面泡沫。
3 用筛滤去汤渣，把勺子放在汤渣上，压出多余汁，待凉即成。

Ingredients

Fish bone, roughly chopped	100 g
Onion, roughly chopped	100 g
Carrot, roughly chopped	100 g
Celery, roughly chopped	80 g
White wine	200 ml
White peppercorn	10 pcs
Bay leaf	2 pc
Lemon	1/2 pc
Water	3 L

Procedures

1 Place the fish bone with all ingredients in pan and bring to the boil.
2 Simer for 20 minutes and skim the foam on the surface.
3 Strain the stock, press the solid with ladle to extract extra liquid and let cool.

奶白沙司 Cream White Sauce

材料 牛油、面粉各15克，牛奶300毫升。

制作方法

1. 把面粉加进溶化的牛油中，以小火烹煮。
2. 其间不断用打蛋器搅动成白面捞，需1～2分钟。
3. 离火，慢慢加入热牛奶，搅入面捞内混合。
4. 回火煮滚，其间不断搅拌，直至熬煮成浓稠的酱汁即可。

Ingredients

Butter	15 g
Flour	15 g
Milk	300 ml

Procedures

1. Pour flour into the melted butter and simmer.
2. Use egg beater to stir the flour to get paste.
3. Off the fire, add the hot milk and mix into the paste.
4. Heat again to boil and stir until getting the thickened sauce.

荷兰沙司 Dutch Sauce

材料 鸡蛋3个，热水45毫升，无盐牛油175克，柠檬1个，盐、胡椒各少许。

制作方法

1 将牛油用热水隔开溶化。
2 滤出牛油清。
3 取出蛋黄倒入碗中，加入牛油清，顺时针方向搅拌。
4 一边搅打，一边加入柠檬汁，再加入盐和胡椒调味。
5 搅打成乳状，调入胡椒即可。

Ingredients

Egg	3 pcs
Hot water	45 ml
Unsalted butter(clarified)	175 g
Lemmon	1 pc
Salt	A little
Pepper	A little

Procedures

1 Melt the butter with hot water.
2 Filter to get the clarified butter.
3 Put the clarified butter into the egg yolks in the bowl and stir clockwise.
4 Add lemmon as stiring and put the salt and fresh pepper in.
5 Stir until it is milky and add again the fresh pepper in it.

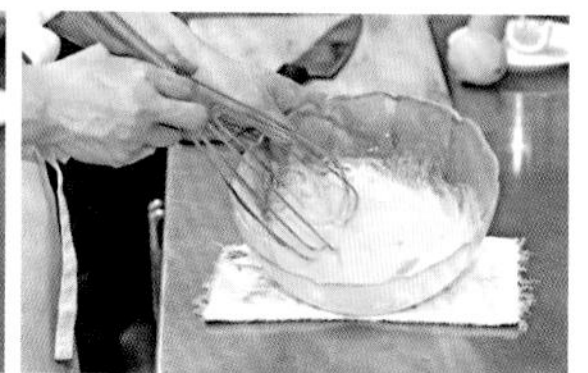

蛋黄酱 Mayonnaise

材料 蛋黄1个，橄榄油、玉米油各150毫升，白醋10毫升，盐、胡椒粉各少许，糖10克，柠檬1个。

制作方法

1 将鸡蛋黄打入碗中。
2 加入盐。
3 加入胡椒粉。
4 加入糖。
5 搅拌均匀。
6 加入橄榄油，不能加太多，否则酱就泻了，朝一个方向搅拌，让蛋黄和油充分吸收。
7 当汁变稠时，加入白醋。
8 加入玉米油。
9 加入柠檬汁即可。

Ingredients

Egg yolk	1 pc	Olive oil, corn oil	150 ml
White vinegar	10 ml	Salt, pepper	A little
sugar	10 g	Lemon	1 pc

Procedures

1 Put the egg yolk into the bowl.
2 Add the salt.
3 Add the pepper.
4 Add the sugar.
5 Stir to make it mixed evenly.
6 Add the olive oil but not too much, otherwise it will be spilled on the sauce. Stir in one direction to make the egg yolk and oil fully absorbed.
7 Add white vinegar when the juice is thick.
8 Add the corn oil.
9 Finally add lemon juice.

千岛汁 Thousand Island Dressing

材料 蛋黄酱500克，西红柿沙司150克，酸黄瓜180克，熟鸡蛋200克，法香20克，洋葱、蒜蓉各10克，白醋10毫升，糖15克，精盐、胡椒粉各少许。

制作方法

1 将蛋黄酱取出，放入碗内。
2 加入切碎的法香。
3 加入切碎的鸡蛋。
4 加入切碎的酸黄瓜。
5 加入蒜蓉。
6 加入切好的洋葱碎。
7 加入西红柿沙司。
8 把所有原料混合拌匀，然后用盐、胡椒粉、白醋、糖调味。
9 搅拌至汁的颜色成粉红色即可。

Ingredients

Mayonnaise	500 g	Onion, chopped	10 g
Tomato sauce	150 g	Garlic, chopped	10 g
Plicked cucumber	180 g	White vinegar	10 ml
Cooked eggs	200 g	Sugar	15 g
Thymus vulgaris	20 g	Salt, pepper	A little

Procedures

1 Take the mayonnaise out into the bowl.
2 Add the salt.
3 Add chopped egg pieces.
4 Add chopped plicked cucumber.
5 Add the garlic.
6 Add chopped onions.
7 Add the tomato sauce.
8 Mix all of the ingredients with salt, pepper, vinegar and sugar seasoning.
9 Wait until the color of the juice turns to pink.

法汁 French Dressing

材料 洋葱碎20克，醋30毫升，橄榄油90毫升，法国芥末酱10毫升，法香碎、蒜蓉各10克，盐、胡椒粉各少许。

制作方法

1 将法国芥末酱放入碗内。
2 加入切好的洋葱碎。
3 加入切好的法香碎。
4 加入蒜蓉拌均匀。
5 加入盐。
6 加入胡椒粉。
7 慢慢加入橄榄油，朝一个方向抽打至浓稠，最后加入白醋即可。

Ingredients

Chopped onion	20 g	Garlic	10 g
Vinegar	30 ml	Chopped thymus vulgaris	10 g
Olive oil	90 ml	Salt, pepper	A little
French mustard	10 ml		

Procedures

1 Take the French mustard into the bowl.
2 Add the chopped onion
3 Add the chopped Thymus vulgaris.
4 Add the garlic. Stir to make it mixed evenly.
5 Add the salt.
6 Add the ground pepper.
7 Add olive oil slowly, Whip in one direction until thickened. Finally add white vinegar.

第三章：头盘类

Appetizers

法式煎鹅肝 French Fried Foie Gras

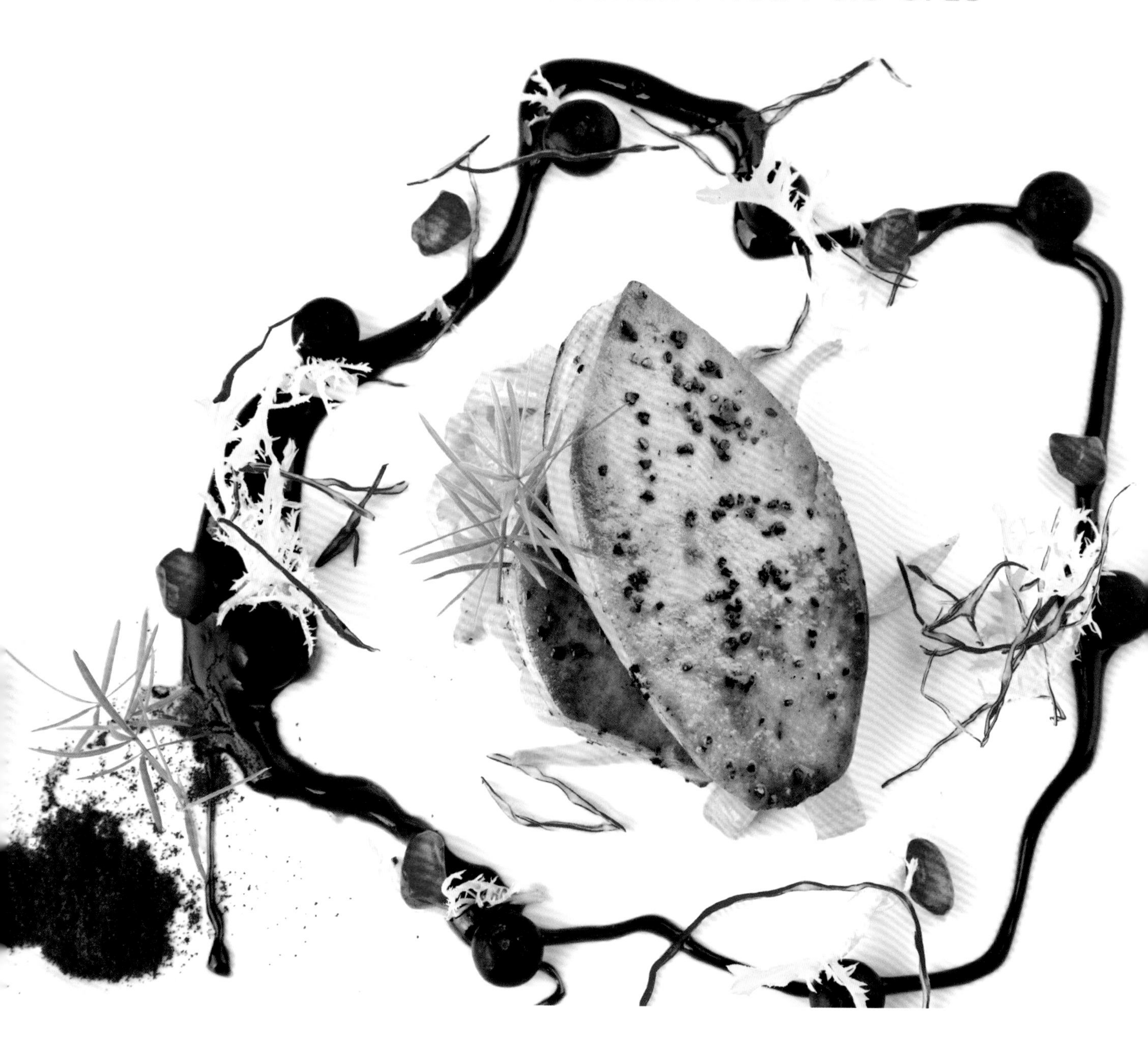

亢　亮 | 行政总厨　沈阳国际皇冠假日酒店

Kang Liang | Executive Chef　Shenyang Crowne Plaza Hotels

西餐专业委员会会长，沈阳市旅游局专家顾问，获得法国蓝带远东区域行政总厨荣誉勋章称号，中国烹饪大师、高级营养师、健康咨询师。

President of Liaoning Western-style Food Professional Committee, Shenyang Tourism Bureau expert consultants, won the fareast region executive chef Medal of honor title from Le Cordon Bleu, China culinary masters, senior nutritionist, health consultant.

材料 鹅肝250克，洋葱40克，苹果1个，蓝莓适量，盐、黑胡椒碎、橄榄油各20克，白兰地15毫升，意大利黑醋35毫升，白糖25克，黄油少许。

制作方法

1 将鹅肝改刀成手指厚片，再加上少许盐、黑胡椒碎和白兰地，腌制20分钟；洋葱去皮，洗净，切丝；苹果洗净，切丁，将洋葱丝和苹果丁混合均匀，用黄油翻炒一下，盛出待用。

2 将意大利黑醋和白糖混合在一起，置火上加热收稠，制成香醋汁。

3 锅置火上烧热，放入适量橄榄油，下入鹅肝，煎至两面呈金黄色。

4 盘中放入用黄油炒过的洋葱丝、苹果丁，再放入煎好的鹅肝，淋上香醋汁，再用新鲜蓝莓装饰即可。

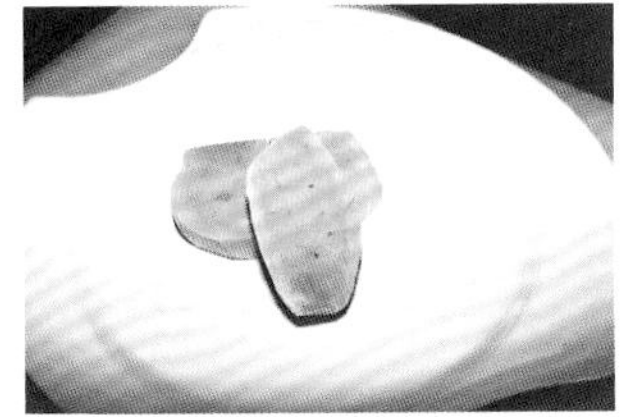

Ingredients

Foie Gras	250 g	Olive oil	20 g
Onion	40 g	Brandy	15 ml
Apple	1 pc	Balsamic vinegar	35 ml
Blueberries	Appropriate	Sugar	25 g
Salt	20 g	Butter	A little
Black pepper	20 g		

Procedures

1 Slice the foie gras to the thickness of a finger, sprinkle a little salt, black pepper and brandy; marinate for 20 minutes, peel onion, wash up, slice and dice, and then mix the apples, onion and diced apples sauté with butter for a moment.

2 Mix balsamic vinegar and sugar, heat to be thickened to make vinegar sause.

3 Put onto the pan, add appropriate amount of olive oil into foie gras, and fry until both sides turn golden yellow.

4 Add onion with butter and diced apples, add fried foie gras, and pour vinegar sauce, and decorated with fresh blueberries.

法式鹅肝酱 French Foie Gras Paste

材料 鹅肝批200克，桃、苹果各1个，蛋黄酱15克，吐司3片，石榴20克，盐、黑胡椒碎各少许。

Ingredients

Foie Gras	200 g
Apple	1 pc
Peach	1 pc
Mayonnaise	15 g
Toast	3 pcs
Pomegranate	20 g
Salt	A little
Black peppercorn	A little

制作方法

1. 将鹅肝批切片；桃去皮，切角；苹果洗净，切条；石榴去壳备用。
2. 混合苹果条、蛋黄酱、盐和黑胡椒碎制成苹果沙拉。
3. 吐司切片，两面烤上色备用。
4. 摆盘，淋上苹果沙拉即可。

Procedures

1. Slice the Foie gras, peel the orange, cut corners, wash the apples, cut it, peel the pomegranate.
2. A mixture of apple，mayonnaise，salt and black peppercorn to make apple salad.
3. Two slice of toast baked colored both sides.
4. Disk up with apple salad.

丁建军 | 副院长　辽宁现代服务职业技术学院
Ding Jianjun | Deputy Dean　Liaoning Modern Service Career Technical College

西餐专业委员会副会长，中烹高级技师，世界中国烹饪联合会教育分会委员，国家级西餐评委。

Vice President of Western-style Food Professional Committee, senior technician of Chinese dishes, member of the World-China culinary federation education, national western food judge.

地中海海鲜沙拉 Mediterranean Seafood Salad

材料 青虾、扇贝、青口贝各3只，土豆、鸡蛋、番茄各1个，荷兰豆40克，意大利黑醋、橄榄油各20毫升，香葱10克，百里香适量。

制作方法

1. 将青虾、扇贝和青口贝清理干净，焯水，冷却备用。
2. 土豆、荷兰豆、鸡蛋洗净，分别煮熟，冷却备用。
3. 番茄洗净，切成块。
4. 混合黑醋、橄榄油、香葱碎、百里香制成沙拉汁。
5. 把所有原料整齐地码放在盘中，淋上沙拉汁即可。

Ingredients

Freshwater shimp	3 pcs	Sweet broad pea	40 g
Mussel	3 pcs	Balsamic vinegar	20 ml
Pectinid	3 pcs	Olive oil	20 ml
Potato	1 pc	Chive	10 g
Egg	1 pc	Thyme	Appropriate
Tomato	1 pc		

Procedures

1. Clean up the freshwater shrimp, pectinid and mussel, scald and cool to reserve.
2. Clean up the potato, sweet broad pea, egg, and cook thoroughly apart, cooling to reserve.
3. Clean up the tomato, cut into pieces.
4. Make salad dressing with balsamic, olive oil, pieces of the chive hyme.
5. Arrange all the ingredients on the plate, drizzle salad dressing.

李　季 | 行政总厨　沈阳市碧桂园玛丽蒂姆酒店

Li Ji | Leader Chef　Country Garden Maritim Hotels in Shenyang

西餐专业委员会副会长，从业18年，在香格里拉集团酒店、洲际集团酒店、玛丽蒂姆集团酒店任职，精通西餐及东南亚菜品。

Vice President of Western-style Food Professional Committee, worked for 18 years, worked in Shangri-La Hotel, InterContinental Hotels Group and Maritim Hotel, proficient in western and Southeast Asian dishes.

凯撒沙拉 Caesar Salad

材料 西生菜200克，烟熏三文鱼3片，芝士棒1个，炒面包丁20克，芝士片、熟培根碎各10克，凯撒汁30克。

制作方法

1 将西生菜洗净，撕成小块。
2 用凯撒汁将生菜拌匀，装入盘中。
3 烟熏三文鱼卷成玫瑰花。
4 摆上芝士棒，撒上面包丁、芝士片、培根碎即可。

Ingredients

Lettuce romaine	200 g
Smoked salmon	3 pcs
Cheese bar	1 pc
Fired bread	20 g
Cooked bacon broken	10 g
Cheese slices	10 g
Caesar dressing	30 g

Procedures

1 Clean up the Lettuce romaine, tear into small pieces.
2 Mix Lettuce romaine with Caesar dressing, dish up.
3 Roll Smoked Salmon into the shape of rose.
4 Place cheese bar,and sprinkle with Fired bread, cheese slices and Cooked Bacon broken.

彭　刚 | 沈阳龙之梦大酒店

Peng Gang | Shenyang Longemont Hotel

西餐专业委员会副会长、西餐高级烹调师。1996年开始从事西餐烹调工作至今，曾任职于香格里拉酒店集团、万豪酒店集团、洲际酒店集团、凯宾斯基酒店集团和喜达屋酒店集团。擅长法式菜肴、德式传统菜肴、东南亚菜肴制作。

Vice President of Western-style Food Professional Committee, senior western food chef. Since 1996 engaged in western food, once worked in Shangri-La Hotel, Wanhao Hotel, InterContinental Hotels Group, Kepinski Hotel, Starwood Hotel. Skilled in French , German traditional and Southeast Asian dishes.

华都夫沙拉 Waldorf Salad

材料 青苹果1个，西芹50克，烤核桃仁30克，蛋黄酱30克。

Ingredients

Green apple	1 pc
Celery	50 g
Roasted walnuts	30 g
Mayonnaise	30 g

制作方法

1 将青苹果洗净，去皮、去核，切条。

2 西芹去叶，洗净，切条，焯水后备用。

3 将苹果条和西芹条用蛋黄酱拌匀。

4 装盘后撒上核桃仁即可。

Procedures

1 Clean up green apple, peel and remove the cores, cut in strips.

2 Cut off leaves of celery, clean up, scald, backup.

3 Mix apple strip and celery strip with mayonnaise.

4 Arrange all the ingredients on the plate, drizzle roasted walnuts.

刘 鹤

Liu He

从业20余年，擅长法国菜、意大利菜，对德餐也颇有心得。

Working in this trade for more than twenty years, skilled in French, Italian dishes, know German dishes clearly and deeply.

金枪鱼意面沙拉 Tuna Pasta Salad

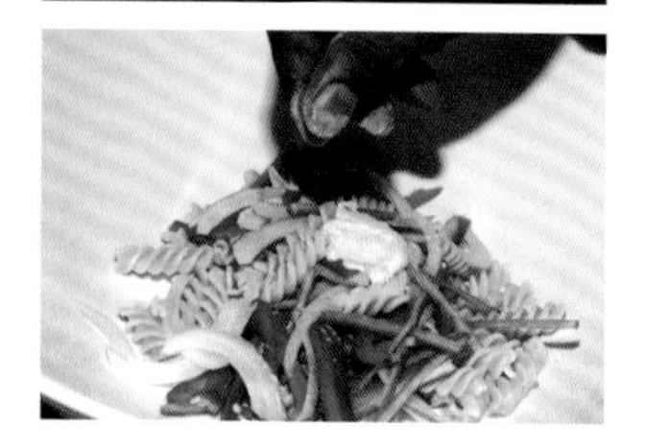

材料 金枪鱼100克，意大利面70克，洋葱、青红圆椒各1个，黑橄榄5个，法香适量，盐、黑胡椒碎各少许，油醋汁30毫升。

制作方法

1 将意大利面煮熟后捞出冷却备用。

2 洋葱、青红圆椒洗净，切丝备用。

3 金枪鱼用盐、黑胡椒碎腌制后烤制成半熟。

4 把煮好的意面、洋葱丝、青红圆椒丝用油醋汁拌匀。

5 把拌好的意面沙拉装入盘中，上面码放好金枪鱼、黑橄榄、法香装饰即可。

Ingredients

Tunas	100 g	Black olive	5 pcs
Pastas	70 g	Parsley	Appropriate
Onion	1 pc	Salt	A little
Red pepper	1 pc	Black pepper	A little
Green pepper	1 pc	Cooking oil vinegar	30 ml

Procedures

1 Bring pastas to the boil and let cool to reserve.

2 Wash onion, red pepper, green peppers and shred to reserve.

3 Souse tunas with salt and black pepper, roast them to the half-cooked.

4 Mix up the pastas, shred onions and peppers with cooking oil vinegar.

5 Put mixed pasta salad on the dishes, decorate with tunas, black olives and parsleys.

孙　强 | 沈阳香格里拉大酒店
Sun Qiang | Shengyang Shangri-La Hotel

西餐专业委员会名誉副会长，沈阳香格里拉大酒店行政副总厨，擅长东南亚美食，精通意式、美式等烹饪，爱好分子料理。

Vice President of Western-style Food Professional Committee, vice leader chef of Shenyang Shangri-La Hotels, skilled in Southeast Asian , Italian and American dishes, interest in Molecular cuisine.

牛油果大虾沙拉 Shrimp and Avocado Salad

材料 大明虾、牛油果各1个，蟹子20克，生菜少许，黄油、盐各少许。

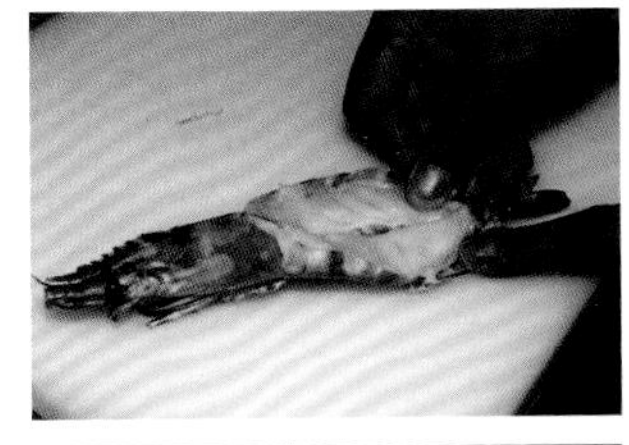

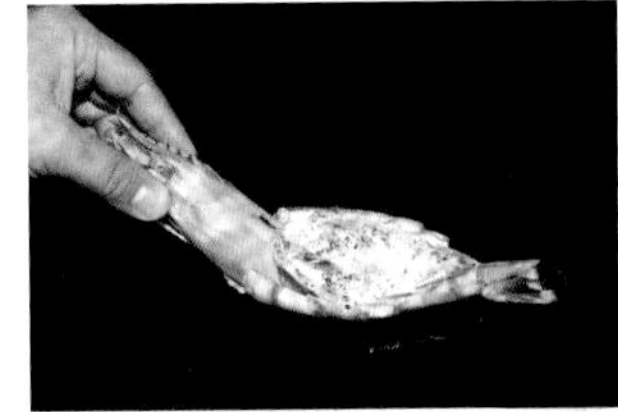

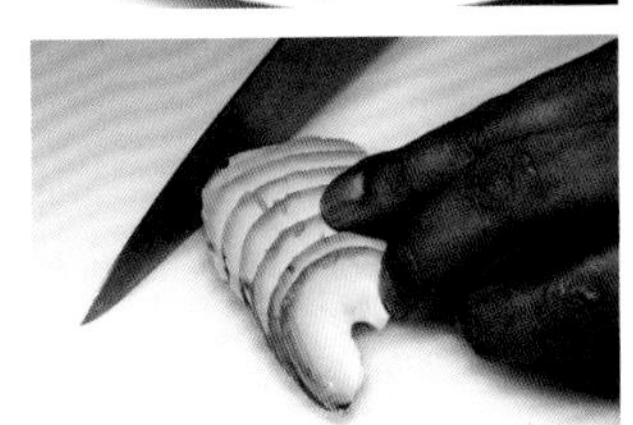

制作方法

1. 大虾留头和尾，开背去壳、去虾线，洗净，用盐腌一会儿。
2. 将腌过的大虾用黄油煎熟备用。
3. 用生菜做盘头装饰。
4. 牛油果去皮、去核，切片，码入盘中，上面放上煎好的大虾。
5. 用蟹子装饰即可。

Ingredients

Prawn	1 pc	Lettuces	A little
Avocado	1 pc	Butter	A little
Crabs	20 g	Salt	A little

Procedures

1. Leave head and tail, shell prawn, devein and rub with salt to remove slime, rinse and pat dry.
2. Bring prawn to the boil with butter, in reserve.
3. Decorate with lettuces.
4. Remove avocado's peel and core, cut into slices, and put prawn on it.
5. Decorate with crab.

张　平 | 莱星顿酒店
Zhang Ping | Vantage Hotel

西餐专业委员会副会长，莱星顿酒店行政总厨，2014年得到辽宁省酒店行业协会及辽沈同行的认可，被评为2014年辽宁省十佳行政总厨。

Vice President of Western-style Food Professional Committee, leader chef of Vantage Hotel, accepted by Liaoning Hotel Committee, awarded 10 top chefs in Liaoning.

德式土豆沙拉 German potato salad

材料 土豆150克，培根末30克，洋葱末20克，香葱10克，白酒醋25毫升，橄榄油15毫升，法式芥末10克，鸡汤、盐、白胡椒粉各少许。

制作方法

1 将土豆煮熟，切片。

2 香葱洗净，切碎。

3 锅中加入鸡汤、白酒醋、法式芥末、培根末、洋葱末、盐、白胡椒粉、葱末，煮开调味。

4 放入切好的土豆，将汁水收稠，取出放凉。

5 出锅摆盘，淋上橄榄油即可。

Ingredients

Potatoes	150 g
Bacon bits	30 g
Onion bits	20 g
Chives	10 g
White wine vinegar	25 ml
Olive oil	15 ml
French mustard	10 g
Chicken soup	A little
Salt	A little
White pepper	A little

Procedures

1 Being potatoes to the boil, cut into slices.

2 Wash chives, cut into bits.

3 Put chicken soup, white wine vinegar, French mustard, bacon bits, onion bits, salt, white peppers, chopped onions into the pot and boil it.

4 Put potato slices into it, sauce thickened, cool it.

5 Dish up and drizzle with olive oil.

马海波 | 行政副总厨　沈阳丽都索菲特酒店

Ma Haibo | Executive Deputy Chef　Shenyang Sofitel Lidu Hotel

拥有10余年五星酒店工作和管理经验，先后在喜达屋及雅高酒店管理集团旗下任职，法国蓝带协会会员，精通法国菜、墨西哥菜、南美菜、东南亚菜等。

Have more than 10 years of working and management experience in five stars hotels, worked in Sheraton Hotels and Accor Hotels respectively, member of Le Cordon Bleu Association, skilled in French, Mexican, South American, Southeast Asian dishes.

樱桃鹅肝 Cherry Foie Gras

安 阳 | 厨师长 沈阳国际皇冠假日酒店

An Yang | Chef Shenyang Crowne Plaza Hotels

法国蓝带美食协会高级会员、西餐高级烹调师。1999年从事西餐工作至今，师从意大利米其林大师法比奥·乌格莱蒂。擅长意大利菜、西班牙先锋分子料理，旁通东南亚菜、法国菜、墨西哥菜等。

Senior member of Le Cordon Bleu, senior western food chef. Since 1999 engaged in western food, learn from Michelin chef Beal Uglady. Skilled in Italian dishes and Spanish Avant-garde Molecular gastronomy, and on general Southeast Asian, French, Mexican dishes.

材料 鹅肝酱150克，红菜汁200克，卡拉胶5克，培根碎、细香葱、樱桃把各少许。

制作方法

1. 将鹅肝酱放入磨具中冷冻。
2. 混合红菜汁和卡拉胶。
3. 将冷冻好的鹅肝球反复在红菜汁中滚蘸，直至菜汁均匀包裹在鹅肝球表面上。
4. 在鹅肝球上插上樱桃把。
5. 装盘时撒上细香葱和炒过的培根碎即可。

Ingredients

Foie gras	150 g
Beetroot sauce	200 g
Carrageenan	5 g
Bacon bits	A little
Chive	A little
Cherry stalks	A little

Procedures

1. Put foie gras into mold, freeze.
2. Mix beetroot sauce and carrageenan.
3. Put the frozen foie gras balls roll in the beetroot sauce, till smooth in the surface.
4. Plug cherry stalks in the foie gras balls.
5. Sprinkle on some chive and sauté bacon bits and dish up.

三文鱼鳄梨卷 Salmon Crocodile Rolls

材料 三文鱼200克，牛油果、柠檬各1个，奶油芝士50克，盐、白糖、白胡椒粉各少许。

制作方法

1. 将三文鱼片成大长片备用，牛油果切丁。
2. 柠檬洗净，取汁，加入白糖搅匀，再放入奶油芝士、盐和白胡椒粉混合均匀，制成芝士酱。
3. 准备一张保鲜膜，将三文鱼铺在保鲜膜上，再均匀地抹上调制好的芝士酱。
4. 在鱼片中心放上牛油果丁，再将三文鱼卷成卷，两头的保鲜膜扎紧，放入冰箱冷冻1小时定形。
5. 将鱼卷改刀成适合的形状，装盘即可。

Ingredients

Salmon	200 g
Avocado	1 pc
Lemon	1 pc
Cream cheese	50 g
Salt	A little
Sugar	A little
White pepper	A little

Procedures

1. Slice salmon and dice avocado.
2. Clean the lemon, squeeze the lemon and mix with sugar, and then mix with cream cheese, salt and white pepper.
3. Spread the salmon on the plastic wrap, and then plastered with cheese sauce.
4. Put diced avocado in the center of salmon sliced and roll it, tied both sides, freeze an hour in the refrigerator.
5. Cave salmon sliced and dish up.

姚国庆 | 副会长　西餐专业委员会

Yao Guoqing | Vice President　Western-style Food Professional Committee

1996年入行，18年国际五星级酒店从业经验，法国蓝带国际美食联合会高级会员，主修法、德、意大利餐，兼修东南亚餐及中餐。

Engaged in western food since 1996, have more than 18 years of working experience in international five stars hotels, senior member of Le Cordon Bleu International Delicacy Association, skilled in French, German, Italian dishes, can do Southeast Asian and Chinese dishes.

烤南瓜沙拉 Baked Pumpkin Salad

田　达 | 行政总厨　成都雅居乐豪生大酒店
Tian Da | Executive Chef　Howard Johnson Agile Plaza in Chengdu

西餐高级烹调师，曾任职于洲际酒店集团、香格里拉集团、豪生集团。专注于西餐烹调事业多年，擅长法国菜、意大利菜、美式、俄式菜等多种西餐的制作。

Senior western cook, once worked in InterContinental Hotels Group, Shangri-La Group, and Howard Johnson Group. Concentrate on western food many years, skilled in French, Italian, American and Russian dishes.

材料 南瓜200克，红圆椒1/2个，柠檬1个，葡萄干15克，盐、黑胡椒碎各少许，柠檬汁25毫升，橄榄油15克。

制作方法

1. 将南瓜去皮、去籽，切成3厘米大的块。
2. 将红圆椒洗净，去蒂及籽，切丁备用。
3. 南瓜用盐、黑胡椒碎、橄榄油搅拌均匀，放入180℃的烤箱烤20分钟。
4. 混合柠檬汁、橄榄油、盐、黑胡椒碎制成沙拉汁。
5. 混合烤好的南瓜块、红椒丁、葡萄干，用沙拉汁拌匀即可。

Ingredients

Pumpkin	200 g	Salt	A little
Bell pepper red	1/2 pc	Black peppercorn	A little
Lemon	1 pc	Lemonade	25 ml
Raisin	15 g	Olive oil	15 g

Procedures

1. Peel the pumpkin, remove the seeds, cut them into 3 cm.
2. Clean the bell pepper red put out the pedicle and seed, dicing.
3. Stir pumpkin with salt, black peppercorn and olive oil evenly. Put them in 180℃ oven 20 minutes.
4. Mix lemonade, olive oil, salt and black peppercorn up into salad dressing.
5. Put the baked pumpkin pieces, bell pepper red and raisin together, stir them with salad dressing.

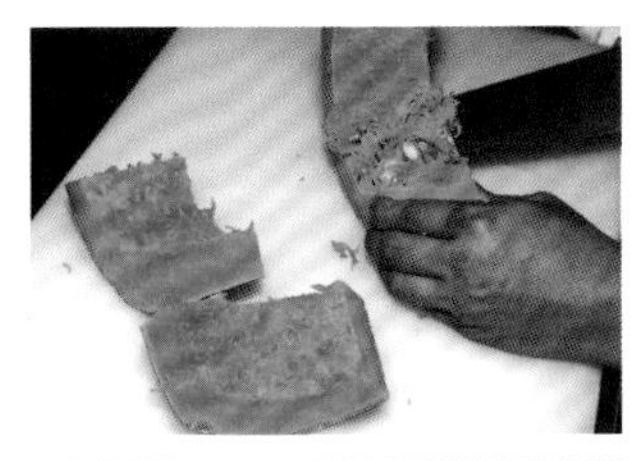
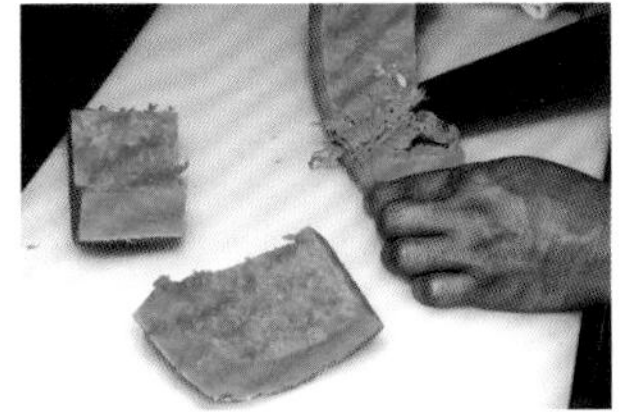

三色果味鱼子 Three Color Fruity Fish Roe

材料 3种果汁各200毫升，钙粉8克，海藻胶15克。

Ingredients

Three kinds of juice	200 ml
Calcium	8 g
Seaweed glue	15 g

制作方法

1 将3种果汁分别混合海藻胶，加热后冷却。
2 将钙粉和清水混合均匀。
3 将3种果汁分别滴入钙水中，形成鱼子状。
4 捞出鱼子，在清水中浸一下即可。

Procedures

1 Mix the three kinds of juice up with seaweed glue, heat and chilling.
2 Mix calcium and water.
3 Drop three kinds of juice in calcium water, put them into the fish roe shape.
4 Fish out the fish roe, put them in water again.

李　杨 | 副会长　西餐专业委员会
Li Yang | Vice President　Western-style Food Professional Committee

曾就职于沈阳洲际假日酒店、沈阳凯宾斯基酒店、丹东皇冠假日酒店、沈阳中山皇冠假日酒店。擅长德国菜和意大利菜。

Once worked in InterContinental Hotels Group, Kaminski Hotel Shenyang, Crowne Plaza Hotels Dandong and Crowne Plaza Hotels Zhongshan Shenyang. Skilled in German, Italian dishes.

烤鸡胸沙拉 Baked Chicken Breast Salad

材料 鸡胸1块，洋葱、菲达芝士各30克，番茄1个，黄瓜20克，柠檬、混合生菜各适量，盐、胡椒粉各少许。

制作方法

1. 鸡胸放入碗中，加入盐、胡椒粉拌匀，腌20分钟。
2. 将腌好的鸡胸放入烤箱内烤制8分钟。
3. 取出熟鸡胸，切片备用。
4. 番茄洗净，切块；黄瓜洗净，切片；混合生菜、芝士片、鸡胸片码放在盘内。
5. 将洋葱碎、柠檬汁、橄榄油、盐、胡椒粉混合均匀，制成沙拉汁。
6. 将沙拉汁浇在摆好的沙拉上即可。

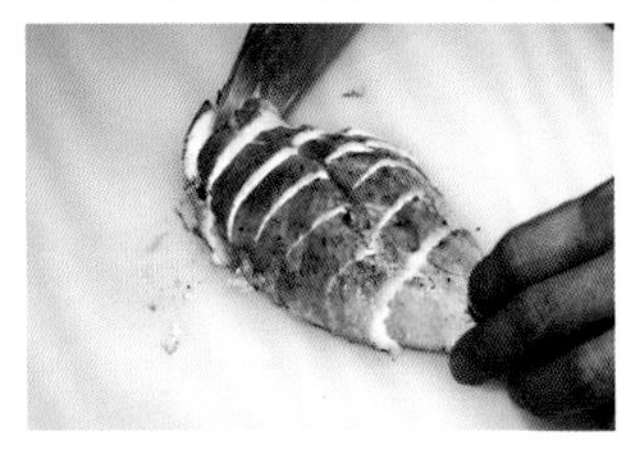

Ingredients

Chicken breast	1 pc	Lemon juice	Appropriate
Minced onion	30 g	Lettuce masculine	Appropriate
Feta	30 g	Salt	A little
Tomato	1 pc	Pepper	A little
Cucumber	20 g		

Procedures

1. Place chicken breasts in a bowl, add salt and pepper mixing, pickle 20 minutes.
2. Place pickled chicken breasts into the oven and bake for 8 minutes.
3. Removed the baked chicken breasts, cut into slices for waiting.
4. Wash tomatoes, cut them into cubes. Wash cucumbers, cut them into slices. Mixed up lettuce masculine, cheese, chicken slices and arrange them on the plate.
5. Mix up minced onion, lemon juice, olive oil, salt, pepper to manufacture salad dressing.
6. Put the salad dressing sauce over the arranged salad.

张岩松 | 高级讲师　沈阳市烹饪职业高中
Zhang Yansong | Senior Lecturer　Shenyang Culinary Vocational Senior High School

1987年任沈阳市烹饪职业高中专职教师至今。

From 1987 to the present, as a full time in the Shenyang Culinary Vocational Senior High School.

泰式牛肉沙拉 Thai Beef Salad

司连福 | 讲师　辽宁现代服务职业技术学院

Si Lianfu | Lecturer　Liaoning Vocational Technical College of Modern Service

西餐专业委员会会员，中式烹调师高级技师，全国餐饮业“中式烹调师”工种二级评委，精通辽菜、粤菜、川菜制作。

Member of Western-style Food Professional Committee, Chinese cooking division senior technician, second-grade judges of nation dining profession "Chinese cooking division" examination, skilled in Liaoning dishes, Cantonese and Sichuan cuisine.

材料 牛柳500克，红洋葱1/2个，黄瓜40克，樱桃番茄20克，紫苏叶3片，香菜20克，罗勒叶3片，姜、蒜各10克，小米辣5克，烤花生适量，青柠汁、芝麻油各15毫升，鱼露10毫升，酱油10克，黄糖5克。

制作方法

1 将姜、蒜切碎；洋葱切丝；樱桃番茄对开；黄瓜切薄片；小米辣切圈；烤花生拍碎；香菜切段。

2 混合青柠汁、黄糖、酱油、芝麻油、鱼露、姜末、蒜末成汁水。

3 将牛柳用调好的汁水腌制两小时。

4 将牛柳烤熟后放置10分钟。

5 将樱桃番茄、黄瓜片、洋葱丝、小米辣圈、香菜段、罗勒叶、花生碎放入容器中。

6 牛柳切薄片，盖在混合好的蔬菜上，浇上调制好的汁水即可。

Ingredients

Beef tenderloin	500 g
Red onion	1/2 pc
Cucumber	40 g
Cherry tomato	20 g
Folia perillae acutae	3 pcs
Coriander	20 g
Basil leave	3 pcs
Ginger	10 g
Garlic	10 g
Millet pepper	5 g
Roasted peanut	Appropriate
Kaffic lime	15 ml
Sesame oil	15 ml
Fish sauce	10 ml
Soy sauce	10 g
Brown sugar	5 g

Procedures

1 Chop up ginger and garlic. Chop onion into thin shreds. Cut cherry tomatoes in half. Cut cucumbers into slices. Cut millet peppers into circles. Break roasted peanut to pieces. Cut coriander into segments.

2 Mix up kaffic lime, brown sugar, soy sauce, sesame oil, fish sauce, chopped ginger, chopped garlic to manufacture juice.

3 Pickle beef tenderloin with the juice for two hours.

4 Bake beef tenderloin and then place them for ten minutes.

5 Place tomatoes, cucumber pieces, onion shreds, millet peppers, coriander segments, basil leave, peanut pieces into container.

6 Cut beef tenderloin into slices, and put them over composite vegetables, pour over modulation juice.

熏三文鱼沙拉 Smoked Salmon Salad

曹 阳 西餐冷菜厨师长 沈阳国际皇冠假日酒店

Cao Yang Leader Chef of Western Cold Appetizer Shenyang Crowne Plaza Hotels

西餐专业委员会委员，西餐高级烹调师。2000年从事西餐工作至今，曾任职于洲际酒店集团、沈阳皇朝万豪酒店、莱星顿酒店，擅长法餐、意大利餐和东南亚餐的烹调制作。

Member of Western-style Food Professional Committee, senior western food chef, engaged in western food since 2000, once worked in International Hotel Corp, Shenyang Empire Wanhao Hotel and Lexington Hotel, skilled in French, Italian and Southeast Asian dishes.

材料 烟熏三文鱼3片，芦笋100克，什锦生菜叶50克，樱桃番茄5个，芝士、黄瓜各20克，盐、胡椒粉各少许，橄榄油15克，香脂醋25毫升。

Ingredients

Smoked salmon	3 pcs
Asparagus	100 g
Lettuce	50 g
Cherry tomato	5 pcs
Cheese	20 g
Cucumber	20 g
Salt	A little
Pepper	A little
Olive oil	15 g
Balsamic vinegar	25 ml

制作方法

1 芦笋洗净，切段，放入沸水中焯烫片刻，捞出过凉。

2 黄瓜洗净，切片；樱桃番茄洗净，切半；芝士切薄片。

3 将橄榄油、香脂醋、盐、胡椒粉放入碗中，制作成沙拉汁。

4 将烟熏三文鱼卷成花状。

5 将蔬菜和三文鱼花摆放在盘中，浇上沙拉汁即可。

Procedures

1 Wash asparagus, cut them into segments and put them blanch in the boiling water for a while, remove them and rinse under running cool water.

2 Wash cucumbers and cut into slices. Wash cherry tomatoes, and cut them in half. Cut cheese into slices.

3 Place olive oil, balsamic vinegar, salt and pepper in a bowl to manufacture salad dressing.

4 Smoked salmon roll into flower.

5 Arrange vegetables and salmon on the plate, put the salad dressing sauce over the plate.

芦笋西柚沙拉 Asparagus Grapefruit Salad

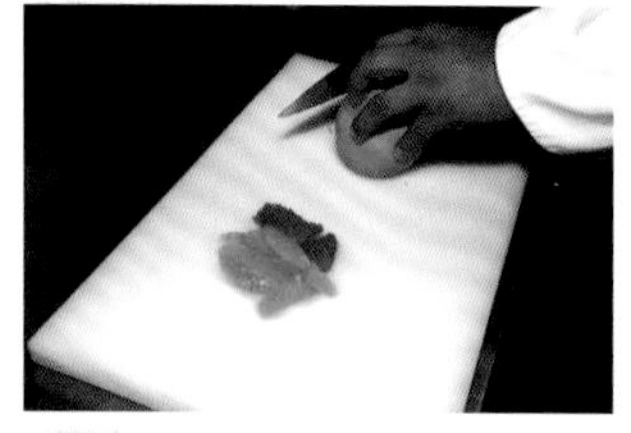
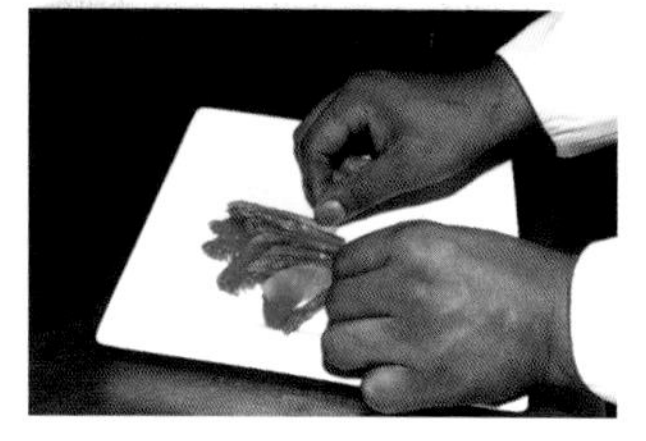

材料 芦笋200克，西柚1个，什锦生菜30克，豆苗15克，酸奶1个。

制作方法

1 将芦笋放入开水中烫熟，再迅速放入冰水中冷却，捞出沥干水分。

2 将什锦生菜、豆苗洗净，切好备用。

3 将什锦生菜放入碗中，加入适量酸奶拌匀。

4 西柚去皮，切角。

5 将芦笋、西柚角码放在盘中，放上拌好的什锦生菜即可。

Ingredients

Asparagus	200 g	Bean seedling	15 g
Grapefruit	1 pc	Yogurt	1 pc
Assorted lettuce	30 g		

Procedures

1 Put the asparagus in boiling water, and then quickly put into the cold water to cool, remove and drain.

2 Wash the Assorted lettuce and bean seedling, cut well to spare.

3 Put the assorted lettuce in a bowl, add moderate yogurt and mix well.

4 Peal the grapefruit, cut the corner.

5 Put the asparagus and grapefruit corner in the plate, and then put the marinade assorted lettuce on it.

郭东海 | 行政总厨　安徽宿州希尔顿逸林酒店

Guo Donghai | Executive Chef　Doubletree of Hilton Anhui Suzhou.

西餐高级烹饪师，1996年从事西餐工作至今，曾任职于凯莱酒店集团、洲际酒店集团、希尔顿集团。

A senior western cook, engaging in western since 1996, worked for Gloria Hotel group, Hilton Group and InterContinental Hotels Group.

嫩煎金枪鱼 Fry Tunas

材料 金枪鱼柳200克，时令蔬菜适量，柠檬汁、糖各少许，面包糠150克，盐、黑胡椒碎各少许，植物油适量。

制作方法

1 将金枪鱼切成适当形状，加入盐、柠檬汁、糖腌制15分钟。

2 将腌制好的金枪鱼表面裹上黑胡椒碎和面包糠。

3 锅中加油烧至七成热，放入金枪鱼炸10秒钟，捞出沥油。

4 将时令蔬菜加工成橄榄形，搭配金枪鱼装盘即可。

Ingredients

Tunas fillet	200 g
Seasonal vegetables	Appropriate
Lemon juice	A little
Sugar	A little
Breadcrumbs	150 g
Salt	A little
Black peppercorn	A little
Vegetable oil	Appropriate

Procedures

1. Cut the tunas fillet into proper shape, add salt, lemon juice and sugar, and then pickle 15 minutes.
2. Coat black peppercorn and breadcrbs on the pickled tunas.
3. Put oil in the pot and burn to seven mature, put the tunas fillet in it to fried ten minutes, then fish out and drain off.
4. Make the seasonal vegetables in to olive shape, match the tunas fillet, and then put them on a plate.

党宏坤 | 西餐厨师长　北约客维景国际大酒店

Dang Hongkun | Leader Chef　Shenyang Grand MeteoPark North Yoker Hotel

西餐高级烹饪师，1996年开始从事西餐行业工作至今，曾任职于香格里拉酒店集团、豪生酒店集团、喜达屋酒店集团，擅长意大利菜和东南亚菜。

A senior western cook, engaging in western cook since 1996, worked for Shangri-La Hotel Group, Starwood Hotel Group, Howard Johnson Hotel Group, skilled in Italy and Southeast Asian dishes.

西班牙蒜香大虾 Spanish Garlic Prawns

材料 8头大虾6只，时令蔬菜适量，蒜末25克，法香碎5克，黄油15克，红粉、盐、黑胡椒碎各适量。

制作方法

1. 大虾去虾线，从背部切开，拍平；时令蔬菜切配好备用。
2. 混合黄油、法香碎、蒜末、盐、黑胡椒碎，制成香草黄油。
3. 将黄油抹在大虾背上。
4. 将大虾放入200℃烤箱，烤8分钟至成熟，取出。
5. 用时令蔬菜适量装饰即可。

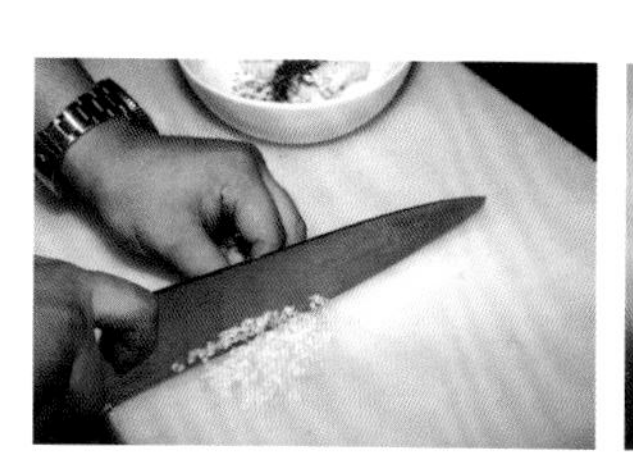

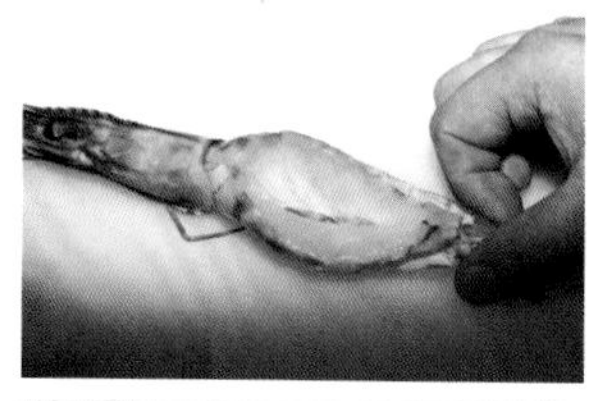

Ingredients

Eight head prawn	6 pcs
Seasonal vegetables	appropriate
Minced garlic	25 g
Ground thymus vulgaris	5 g
Butter	15 g
Paprika	Appropriate
Salt	Appropriate
Black peppercorn	Appropriate

Procedures

1. Devein the prawns, cut from the back, put flat and mix with some seasonal vegetables for later use.
2. Mix the butter, Ground thymus vulgaris, minced garlic, salt and black peppercorn to make herb butter.
3. Spread the butter over the back of prawns.
4. Roast the prawns in the oven at 200℃ for 8 minutes and take out.
5. Decarate with the seasonal vegetables.

吴 钢

Wu Gang

从业近20年，曾经在柏联集团精品酒店、香格里拉酒店、喜来登酒店等多家中外酒店集团服务，师从荷兰星级米其林大厨Niels Halen，精通于新派法国菜、意大利菜及融合菜。

Close to 20 years of working experience, once serviced in a number of Chinese and foreign hotel groups, such as Bolian Group Boutique Hotel, Shangri-La Hotel and Sheraton Hotel, learned from Holland Michelin star chef Niels Halen, skilled in new French , Italian and Fusion dishes.

墨西哥鸡肉薄饼 Mexico Chicken Pancake

蔡　巍 | 行政总厨　东方银座国际酒店

Cai Wei | Executive Chef　Oriental Ginza International Hotel

西餐专业委员会会员，1996年从事西餐工作至今，曾先后在万豪集团、喜达屋集团、雅高集团工作过，擅长法国菜、意大利菜、东南亚菜的制作。

Member of Western-style Food Professional Committee, engaged in western food since 1996, once worked in Marriott Hotel, Starwood Hotels and Accor Group, skilled in French, Italian and Southeast Asian dishes.

材料 鸡肉丝100克，洋葱、牛油果各30克，彩椒20克，番茄丁10克，蒜末5克，薄饼2张，香菜少许，橄榄油、卡真粉、盐、胡椒粉、糖、柠檬汁、辣椒仔、马苏里拉芝士各适量。

制作方法

1 洋葱、彩椒均切丝备用。
2 锅内加油烧热，放入鸡肉丝、彩椒丝、洋葱丝炒至成熟，加入卡真粉、盐调味，制成鸡肉馅料。
3 混合番茄丁、香菜、蒜末、盐、橄榄油、柠檬汁，制成番茄沙沙。
4 用薄饼夹住鸡肉馅料和芝士。
5 将带馅料的薄饼放入200℃的扒板上煎至两面金黄，切开。
6 盘中放入煎好的薄饼，淋上番茄沙沙即可。

Ingredients

Chicken shreds	100 g
Onion	30 g
Avocado	30 g
Color pepper	20 g
Diced tomatoes	10 g
Garlic	5 g
Pancakes	2 pcs
Caraway	A little
Olive oil	Appropriate
Cajun seasoning	Appropriate
Salt	Appropriate
Pepper	Appropriate
Sugar	Appropriate
Lemon juice	Appropriate
Capsicum	Appropriate
Mozzarella cheese	Appropriate

Procedures

1 Cut the onion and color pepper into shreds.
2 Heat up the pan with oil, place chicken shreds, color pepper shreds and onion shreds in it to cooked, season with Cajun seasoning and salt, make into chicken fillings.
3 Mix diced tomatoes, caraway, garlic, salt, olive oil and lemon juice, make into tomato salsa.
4 Clamp chicken fillings and cheese with pancakes.
5 Put pancakes with fillings on the 200℃ grill until both sides are golden.
6 Put the fried pancakes in the plate, pour over the tomato salsa.

香煎扇贝配芒果酱 Fried Scallop with Mango Sauce

材料 大扇贝2只，芒果、红圆椒各1个，柠檬皮、洋葱碎各少许，白葡萄酒、柠檬汁各25毫升，橄榄油15克，盐、黑胡椒各少许。

制作方法

1 将扇贝肉从壳中取出备用。

2 碗中放入扇贝肉，加入盐、黑胡椒、柠檬皮，腌制10分钟。

3 锅中放入红甜椒、芒果肉、洋葱碎、葡萄酒、盐、糖烧开，出锅用打碎机打成泥。

4 将扇贝肉放入油锅中煎至两面金黄。

5 出锅装盘，适量装饰即可。

Ingredients

Scallop	2 pcs
Mango	1 pc
Red pepper	1 pc
Lemon sliced	A little
Minced onion	A little
White wine	25 ml
Lemon juice	25 ml
Olive oil	15 g
Salt	A little
Black pepper	A little

Procedures

1 Take scallop out of the shell.

2 Put the scallop in the bowl, add salt, black pepper and lemon peel, and souse it 10 minutes.

3 Boil the pot with red pepper, mango, minced onion, wine, salt and sugar, beat into mud with beating crusher.

4 Put the scallop in the frying-pan until both sides are golden.

5 Dish up, decorate it suitably.

钱　丰 | 厨师长　湖州雷迪森阳光假日酒店
Qian Feng | Leader Chef　Huzhou Landison Sun Holiday Hotel

西餐高级烹调师，1997年从事西餐工作至今。曾任职于凯莱酒店集团、香格里拉集团、世贸酒店集团和雷迪森酒店集团，擅长意大利传统菜、法国菜、东南亚菜。

Western senior cook, engaged in western food since 1997. Once worked in Gloria Hotel Group, Shangri-La Group, World Trade Hotel Group and Landison Hotel Group, skilled in Italian traditional dishes, French and Southeast Asian dishes.

炸乡村鸡翅 Country Fried Chicken Wings

陈　颂

Chen Song

从业20余年，在爱尔兰米其林餐厅工作学习多年。擅长日式料理、传统俄式菜肴和爱尔兰菜。

In this trade for more than twenty years, working and studying in Irish Michelin starred restaurants for many years. Skilled in Japanese cuisine, traditional Russian cuisine and Irish dish.

材料 鸡中翅5只，鸡蛋2个，面包糠150克，面粉少许，植物油1升，泰式鸡酱35克，奥尔良调料20克。

制作方法

1 将鸡中翅用奥尔良调料腌制2小时。

2 将鸡蛋打成蛋液。

3 将腌制好的鸡翅拍上面粉，蘸匀鸡蛋液，再裹上面包糠备用。

4 锅中加油烧至七成热，下入鸡翅炸熟，出锅装盘。

5 配上小沙拉和泰式鸡酱即可。

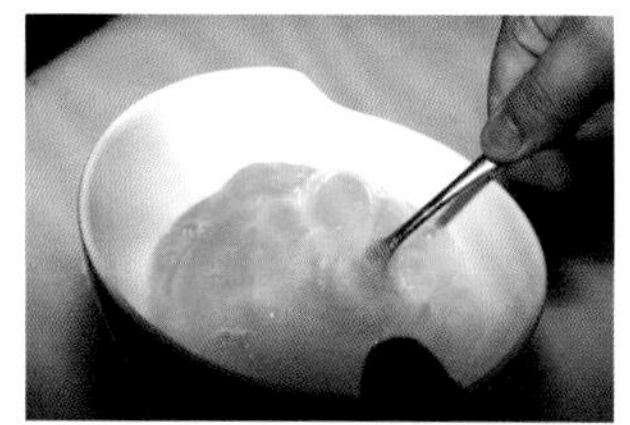

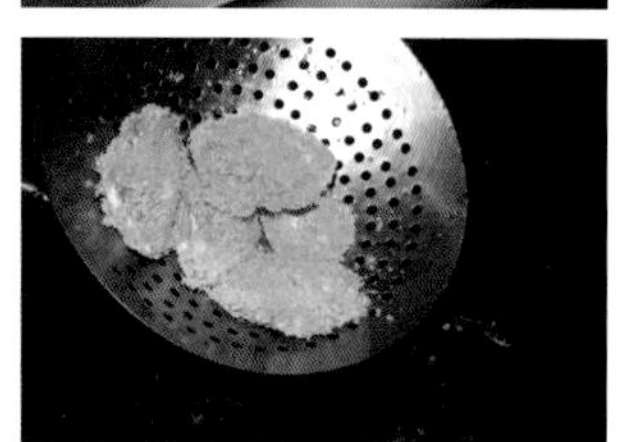

Ingredients

Mid joint chicken wing	5 pcs	Vegetable oil	1 L
Egg	2 pcs	Thai chicken sauce	35 g
Breadcrumbs	150 g	Orlean seasoning	20 g
Flour	A little		

Procedures

1 Leave mid joint chicken wing with Olean seasoning to marinade for two hours.

2 Stir the eggs into a liquid.

3 Make flour click on the marinated chicken wings, evenly dip egg liquid, wrapped in breadcrumbs, spare.

4 Pour oil into the pan, put it to 70% heat, pour the chicken wings, fry them, and transfer the chicken wings and dish.

5 Add some salad and Thai chicken sauce.

东南亚风味鱿鱼圈 Southeast Asian Flavor Squid Ring

材料 鱿鱼1个，咖喱粉、面粉各50克，芝麻10克，鸡蛋液适量，植物油1升，鸡精10克。

Ingredients

Squid	1 pc
Curry powder	50 g
Flour	50 g
Sesame	10 g
Egg liquid	Appropriate
Vegetable oil	1 L
Chicken essence	10 g

制作方法

1. 将清理干净的鱿鱼切圈。
2. 混合咖喱粉、鸡精、面粉、芝麻。
3. 将鱿鱼圈先拍上面粉，再裹上鸡蛋液，然后沾上混合粉。
4. 将制作好的鱿鱼圈在170℃的热油中炸成金黄色。
5. 出锅装盘，适量装饰即可。

Procedures

1. Cut the clean squid into circles.
2. Mix curry powder, chicken essence, flour, sesame.
3. Dredge the squid with flour, wrapped egg liquid, then coat with mixing powder.
4. Fried the ready-made squid rings to golden in 170℃ oil.
5. Remove from the pan and put on a plate, appropriate decoration that will do.

孟 宇 | 西餐厨师长 长白山假日度假酒店

Meng Yu | Leader Chef Holiday Inn Resort in Changbai Mountain

西餐专业委员会副秘书长，中餐高级烹调师，西餐中级烹调师，从业13年，在多家四星、五星级酒店就职，擅长墨西哥菜、意大利菜、东南亚菜、德式菜，在传统菜式的基础上添加自我元素与现代营养膳食搭配，做出不同风格的菜品。

Deputy general secretary of Western-style Food Professional Committee, senior chef of Chinese cooking, intermediate cook of Western food , working for 13 years, once worked in many four-star, five-star hotels, skilled in Mexican cuisine, Italian cuisine, Southeast Asian cuisine, German cuisine, add personal elements on the basis of traditional dishes, paired with modern nutritional diet,make different styles of dishes.

意大利式文蛤贝 Italian Clam Shellfish

材料　大文蛤500克，蒜蓉25克，鲜法香碎5克，白葡萄酒25毫升，淡奶油35克，橄榄油20克。

制作方法

1. 锅中加入橄榄油烧至温热，下入蒜蓉炒成金黄色。
2. 下入处理好的文蛤。
3. 加入适量白葡萄酒煮半分钟。
4. 加入淡奶油至收稠汁水。
5. 加入鲜法香碎，即可出锅装盘。

Ingredients

Large clam shellfish	500 g	White wire	25 ml
Garlic	25 g	Unsalted butter	35 g
Fresh thymus	5 g	Olive oil	20 g

Procedures

1. Add olive oil to the pan and cook until warm, The minced garlic and stir fry until golden brown.
2. Put in the dealt large clam shellfish.
3. Add the right amount of white wine and cook for half a minute.
4. Add the unsalted butter to thickened juice.
5. Add fresh thymus, and put in a pan.

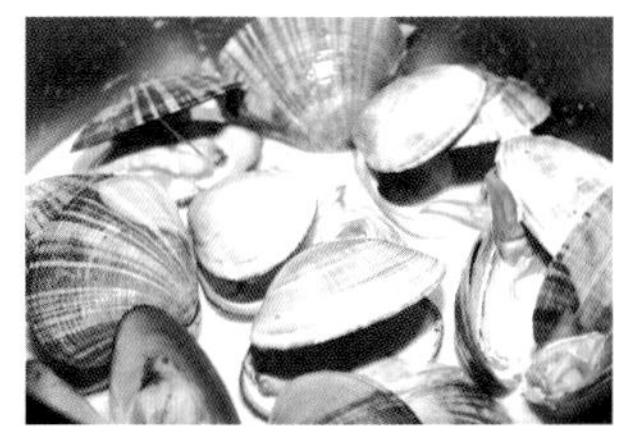

高大伟 | 西餐教师　辽宁现代服务职业技术学院

Gao Dawei | Instructor　Liaoning Vocational Technical College of Modern Service

西餐专业委员会会员，西餐高级烹调师、高级营养师，法国蓝带美食协会荣誉会员，擅长中西创意菜式。

Member of Western-style Food Professional Committee, senior western food cook, senior nutritionist, honored member of Le Cordon Bleu Delicacy Association, specializing in western creative cuisine.

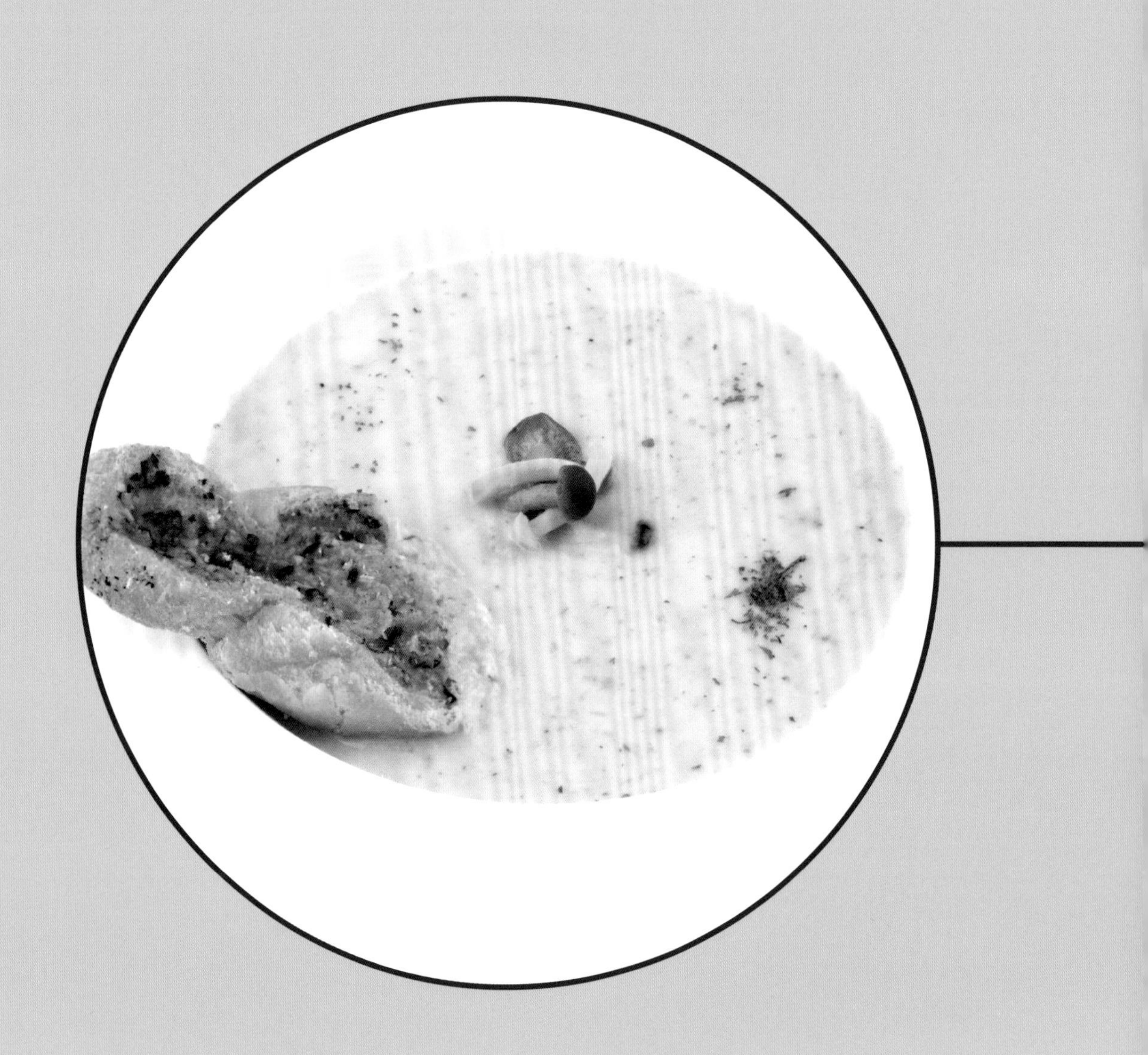

第四章：汤类

Soup

意大利蔬菜汤 Minestrone Soup

吕　放 | 高级主管　沈阳皇冠假日酒店
Lü Fang | Senior Supervisor　Shenyang Crowne Plaza Hotels

从事酒店行业15年，拥有西餐厨师资格证二级品酒资格证，HYGIENE MANAGEMENT SYSTEMS资格证，对西餐制作有丰富的经验，致力于传播纯正的西餐文化和米其林星级的服务。

Working in this trade for 15 years, hold the 2-degree certificate for wine appreciators in Western-style food, certificate of HYGIENE MANAGEMENT SYSTEMS, experienced in Western-style food, devoted to the spread of Western-style food culture and Michelin star service.

材料 各种时令蔬菜500克，意大利面25克，蒜5克，洋葱块少许，橄榄油25毫升，番茄膏20克，盐、芝士粉各10克，什锦香草3克，鸡汤适量。

制作方法

1 将各种蔬菜切小片备用，蒜切成末。

2 锅中加入橄榄油烧热，下入洋葱块炒出香味，再加入蒜末炒香。

3 下入切好的时令蔬菜炒出香味，再加入番茄膏炒2分钟。

4 加入鸡汤、什锦香草、盐煮20分钟。

5 装盘时搭配适量意大利面和芝士粉即可。

Ingredients

A variety of seasonal vegetables	500 g	Tomato paste	20 g
Pasta	25 g	Salt	10 g
Garlic	5 g	Cheese powder	10 g
Onion blocks	A little	Assorted vanilla	3 g
Olive oil	25 ml	Chicken soup	Appropriate

Procedures

1 Cut all kinds of seasonal vegetables into small pieces, garlic cut into the end.

2 Add the olive oil into hot pot with the diced onion to sauté until being fragrant, and add garlic and sauté until fragrant.

3 Put in the chopped vegetables and stir fry until being fragrant, then add tomato paste and stir to santé for 2 minutes.

4 Add chicken soup, salt, assorted vanillas and cook for 20 minutes.

5 With the right amount of pasta and cheese powder.

法式洋葱汤 French Onion Soup

材料 洋葱500克，黄油100克，鸡汤或牛肉汤适量，盐少许。

Ingredients

Onion	500 g
Butter	100 g
Chicken soup or beef soup	Appropriate
Salt	A little

制作方法

1 将洋葱切成细丝。

2 锅中加入黄油烧至温热，下入洋葱丝煸炒片刻。

3 将洋葱丝炒软呈金黄色。

4 加入适量鸡汤或牛肉汤煮20分钟。

5 出锅装盘时搭配芝士焗法包即可。

Procedures

1 Cut the onion into thin filaments.

2 Add butter into the pan and cook until warm, put the onions in and sauté for a while.

3 Stir to sauté the onion until golden yellow.

4 Add the right amount of chicken soup or beef soup for 20 minutes.

5 Place the dish with baked cheese bread collocation.

李晓峰 | 西餐高级主管　沈阳皇冠假日酒店

Li Xiaofeng | Senior Supervisor　Shenyang Crowne Plaza Hotels

西餐专业委员会会员，从事酒店厨房行业18年，精通西餐和中餐面点，法国蓝带美食协会会员，2013年获得《新西餐》专业杂志评选的“年度优秀大厨”称号。

Member of Western-style Food Professional Committee, engaged in the hotel chef industry for 18 years, proficient in Western-style food and Chinese pastry, member of Le Cordon Bleu Delicacy Association, in 2013 won the *New Western-style Food* magazine named “Outstanding Master” title.

马赛海鲜汤 Bouillabaisse

材料 各种新鲜海鲜350克，洋葱、胡萝卜、西芹各30克，番茄2个，藏红花0.5克，柠檬1个，海鲜汤底适量，白葡萄酒15毫升，盐、胡椒粉各少许。

制作方法

1. 将各种新鲜海鲜清理干净，切成丁；各种蔬菜切丁；柠檬切成瓣。
2. 锅中加油烧热，炒香各种蔬菜，再加入白葡萄酒、藏红花和海鲜汤底烧沸。
3. 加入各种海鲜煮至成熟，加入盐和胡椒粉调味。
4. 出锅装盘后搭配柠檬瓣即可。

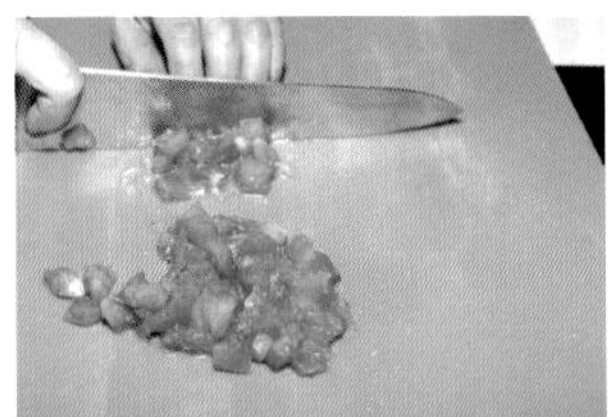

Ingredients

All kinds of fresh seafood	350 g	Saffron	0.5 g
		Lemon	1 pc
Onion	30 g	Seafood soup	Appropriate
Carrot	30 g	Amontillado	15 ml
Celery	30 g	Salt	A little
Tomato	2 pcs	Pepper	A little

Procedures

1. Clean up all kinds of fresh seafood, cut into cubes. All kinds of vegetables diced. Lemon is sliced.
2. Heat oil in a wok, stir fry vegetables, add Amontillado, saffron, seafood soup boil.
3. Adding a variety of seafood cooked to mature, add salt and pepper.
4. A dish with collocation of lemon flap.

刘　冬 | 厨师长　沈阳国际皇冠假日酒店

Liu Dong | Leader Chef　Shenyang Crowne Plaza Hotels

西餐专业委员会会员，从事厨师行业20余年，擅长中式烹调、西式料理及大型宴会等。曾在洲际集团、香格里拉集团、凯莱集团工作过多年，目前兼职酒店食品安全经理。

Member of Western-style Food Professional Committee, engaged in more than 20 years of cook industry, skilled in Chinese cooking, Western food and large banquet, etc. Worked for many years in InterContinental Hotels Group, Shangri-La group, Kailai group has, working part time in hotel as food safety manager now.

匈牙利牛肉汤 Goulash Soup

材料 牛柳250克，洋葱45克，胡萝卜、西芹各30克，蒜末10克，土豆100克，番茄膏20克，红酒30毫升，香草3克，面粉、盐、胡椒粉各少许，甜辣椒粉5克，牛肉汤底适量。

制作方法

1 将洋葱、西芹、胡萝卜和土豆处理干净，均切成丁。

2 牛柳切丁。

3 将牛柳丁放入锅中炒上色，再加入甜辣椒粉炒匀。

4 加入蒜末、番茄膏、面粉、红酒炒2分钟，再加入牛肉汤底。

5 加入各种蔬菜丁，煮至牛柳丁软嫩。

6 加入盐、胡椒粉、香草调好味道即可。

Ingredients

Beef fillet	250 g
Onion	45 g
Carrot	30 g
Celery	30 g
Minced garlic	10 g
Potato	100 g
Tomato paste	20 g
Red wine	30 ml
Vanilla	3 g
Flour	A little
Salt	A little
Pepper	A little
Sweet pepper	5 g
Beef soup	Appropriate

Procedures

1 The onion, celery, carrots and potatoes are cut into small clean.

2 Cut beef fillet into pieces.

3 Add the beef fillet into the pan and stir well, add the sweet pepper and stir well.

4 Add the minced garlic, tomato paste, flour, red wine for 2 minutes, then add the beef soup.

5 Add a variety of vegetables, and cook until the beef fillet is soft and tender.

6 Add salt, pepper, vanilla to keep good taste.

陈 旭

Chen Xu

从事餐饮行业11年以上，包括跟随外籍米其林三星大师学习意大利大餐及经典欧陆菜3年，东南亚菜2年，料理2年。曾就职于洲际酒店集团6年，具有良好的职业道德与操守。

He was engaged in the catering industry for more than 11 years, including learning to follow the foreign master of the Michelin 3-star who master Italian cuisine and classic European dishes for 3 years,and 2 years in Southeast Asian dishes, cooking for 2 years. He was also employed by InterContinental Hotels Group for six years, having a good professional ethic and integrity.

泰式冬荫功汤 Thailand Tom Yang Soup

关　威 | 行政总厨　蓝斯缇纳主题餐厅

Guan Wei | Executive Chef　Lance Twips Themed Restaurant

在欧美游历数十年，曾就职于都柏林最顶级的ELY餐厅。回国后根据国内外饮食上的差异，提出了属于自己的创意厨房概念。在2013年被评为《新西餐》杂志年度优秀大厨。

Traveled in European countries for several decades, who was empolyed by the Dorbolin's top LEY restaurant.Accroding to the domestic and international diffiences in diets, he offered his own concepts about innovation kitchen after returning back to our country. He was prized as the excellect chief of the *New Western-style Food* magazine in 2003.

材料 大虾5只，蘑菇30克，蒜、小米辣各5克，香茅、香菜各10克，樱桃番茄5个，柠檬1/2个，冬荫功酱15克，虾汤底适量。

Ingredients

Prawn	5 pcs
Mushroom	30 g
Garlic	5 g
Capsicum frutescens	5 g
Citronella	10 g
Coriander	10 g
Cherry tomato	5 pcs
Lemon	1/2 pc
Tom yam past	15 g
Shrimp stock	Appropriate

制作方法

1 将大虾清理干净备用。

2 蒜切片；小米辣切末；蘑菇切成片；香菜切成末。

3 锅中加油烧热，下入蒜片、小米辣末、香茅炒出香味，再加入大虾炒上色。

4 加入冬荫功酱炒1分钟，再加入虾汤底、樱桃番茄和蘑菇片煮3分钟。

5 淋上少许柠檬汁，出锅装碗，加入香菜末即可。

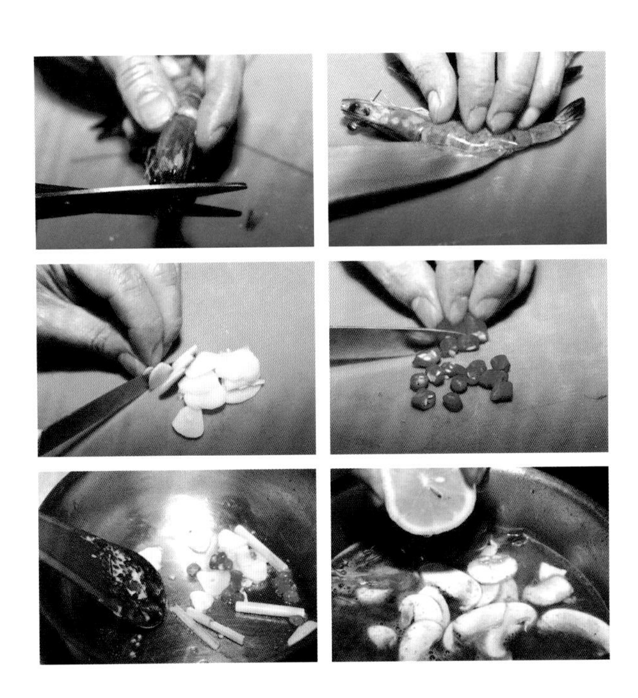

Procedures

1 Wash prawns and set aside.

2 Slice the garlic and mushroom, cut capsicum frutescens and coriander to pieces.

3 Heat the oil in the pan, pour into mashed garlic and capsicum frutescens, citronella, fry fragrant, add into prawn and scambled out of the color.

4 Add Tom yam past to sauté for one minute, and add shrimp stock and cherry tomato and mushroom pieces, cook for 3 minutes.

5 Drizzle with a little lemon juice, add coriander pieces.

番茄汤 Tomato Soup

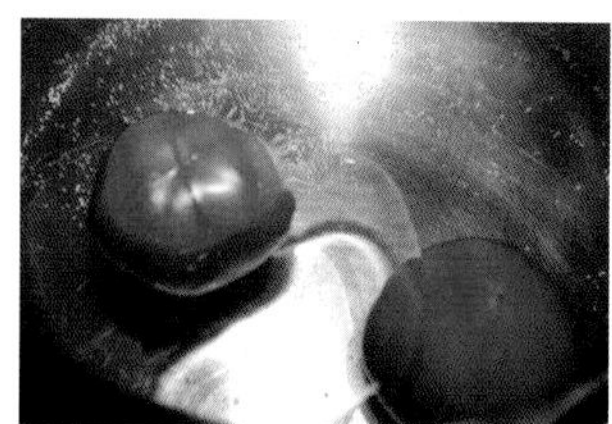
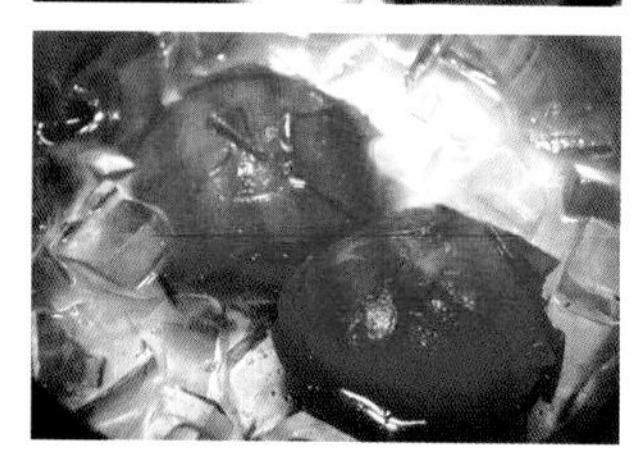
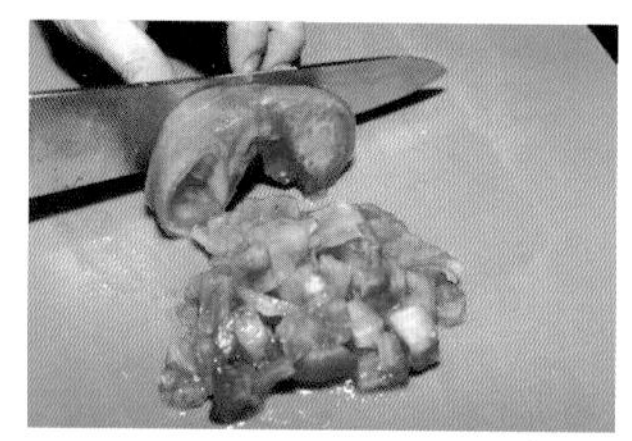

材料 番茄300克，洋葱块50克，土豆块80克，盐、糖各少许，香叶2片，番茄膏20克，黄油30克，鸡汤底适量。

制作方法

1. 将番茄去蒂，洗净，在正面划上十字口。
2. 锅中加水烧沸，下入番茄煮10秒，捞出过凉。
3. 将番茄去皮、去籽，切成小块。
4. 锅中加黄油烧热，下入洋葱块炒软，再加入番茄块和番茄膏略炒。
5. 加入鸡汤底和土豆块煮20分钟，再用打碎机将汤搅拌至黏稠。
6. 将打好的汤二次加热，加入盐、糖调味即可。

Ingredients

Tomato	300 g	Myrcia	2 pcs
Onions pieces	50 g	Tomato paste	20 g
Potato pieces	80 g	Butter	30 g
Salt	A little	Chicken soup	Appropriate
Sugar	A little		

Procedures

1. Remove the head of the tomato，wash clean，create crow ont he front.
2. Boil a pot of water，cook tomato for 10 second，fish out and wait it cold.
3. Remove the tomato skin and seed，cut into slices.
4. Add butter in a hot pan，add onions pieces and fried it to soft，next fried tomato paste and tomato pieces.
5. Add chicken soup and potato pieces cook for 20 minute，then stir the soup to the thick.
6. Twice heat the soup, add salt, sugar seasoning.

刘　贺 | 厨师长　沈阳蓝斯缇纳跨界主题餐厅
Liu He | Head Chef　Shenyang Lance Twips Themed Restaurant

西餐专业委员会会员，从业13年，游历世界各国学习和交流西餐文化，深入了解餐饮精髓，对西餐文化有独到见解和认识，曾任职于多家高档餐厅和酒店，擅长西式意境菜、创新菜以及分子料理。

Member of Western-style Food Professional Committee, employed for 13 years, traveled the world to learn and exchange western food culture, having the special understanding and insights of the essence of dining on western culture. He worked at a number of high-end restaurants and hotels, skilled in western style artistical conception and innovative dishes and molecular gastronomy.

奶油南瓜汤 Cream Pumpkin Soup

材料 南瓜300克，洋葱50克，香叶、盐、胡椒粉各少许，鸡汤底适量。

制作方法

1 将南瓜去皮，切丁。

2 洋葱切丝。

3 锅中加黄油烧热，下入洋葱丝炒出香味，再加入南瓜丁。

4 将南瓜丁炒软并炒出香味。

5 加入香叶和鸡汤底煮至25分钟，再用打碎机搅拌成糊状。

6 将打好的汤二次加热，放入盐、胡椒粉调味即可。

Ingredients

Pumpkin	300 g
Onion	50 g
Myrcia	A little
Salt	A little
Pepper	A little
Chicken soup	Appropriate

Procedures

1. Peel and dice the pumpkin.
2. Cut onions into shreds.
3. Place butter and heat the wok, put in the onions silk to stir until you smell delicious flavor. Add pumpkin patches.
4. Stir pumpkin until melt and smell delicious flavor.
5. Place myrcia and chicken soup, boil until 25 cents change it into paste with blender.
6. Heat twice , mix sauce with salt and pepper.

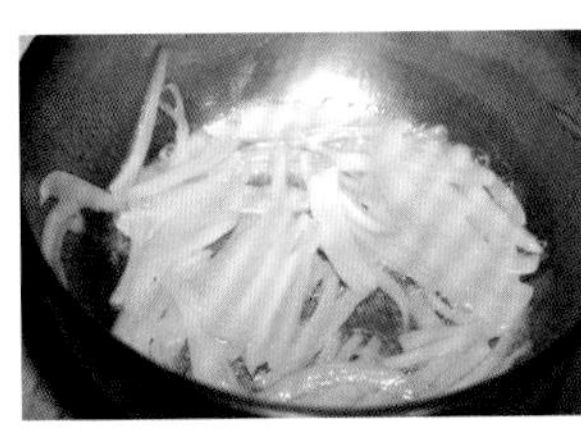

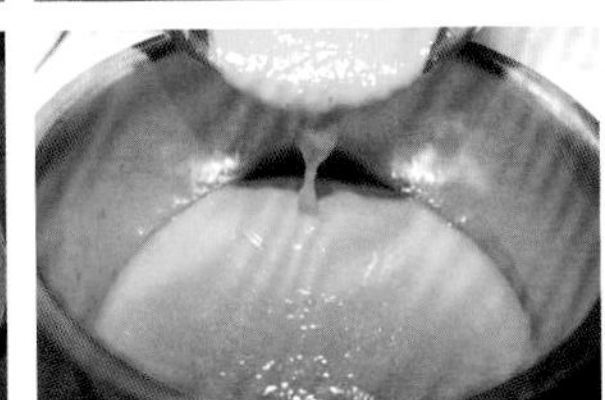

陈　辉 | 厨师长　沈阳丽都索菲特酒店

Chen Hui | Leader Chef　Shenyang Lidu Sofitel Hotel

西餐专业委员会会员，从事餐饮工作17年，先后在格林大酒店、本溪富红酒店等任职，拥有丰富的烹饪经验，精通各种西式美食。

Member of Western-style Food Professional Committee, engaged in catering for 17 years, once worked in Green Hotel and Benxi Fuhong Hotel etc., experienced in cook, skilled in all kinds of western food.

奶油芦笋汤 Cream Asparagus Soup

材料 芦笋250克，洋葱20克，黄油25克，盐、黑胡椒各少许，鸡汤底适量。

制作方法

1 将芦笋洗净，去根，切段。

2 洋葱切丝。

3 锅中加黄油烧热，下入洋葱丝炒出香味，再加入芦笋段炒香。

4 加入鸡汤底煮至芦笋软烂，再用打碎机搅拌成糊状。

5 将打好的汤二次加热，再加入盐、黑胡椒调味即可。

Ingredients

Asparagus	250 g
Onion	20 g
Butter	25 g
Salt	A little
Black pepper	A little
Chicken soup	Appropriate

Procedures

1 Rinse asparagus cut root and cut it into segments.

2 Cut onions into shreds.

3 Place butter and heat the wok, put in the onions silk stir until you smell delicious flavor. Put asparagus segments into wok continue to stir.

4 Place chicken soup and bring to boil until melt. Change it into paste with blender.

5 Heat twice , sauce with salt and black pepper.

种　磊 | 西餐厨师长　沈阳华府酒店
Zhong Lei | Leader Chef　Shenyang Huafu Hotel

西餐高级烹调师，1998年从事西餐工作至今，曾任职于喜达屋集团、美国优势集团和天伦集团，师从于多位西餐烹调大师，擅长意大利菜、德式菜、法国菜、东南亚菜及分子料理。

Senior western food chef. Engaged in western food since 1998, once worked in Starwood Hotels, American advantage Group, Tian Lun Group. Learn from many western food chef, skilled in Italian, German, French, Southeast Asian dishes and Molecular gastronomy.

奶油蘑菇汤 Cream Mushroom Soup

王德义 | 西餐厨师长　沈阳jin酒店

Wang Deyi | Leader Chef　Shenyang Jin Hotel

从业20余年，擅长东南亚菜、传统西餐及中餐，是不可多得的中西双修的优秀厨师长。

Engaged in western food more than twenty years, skilled in Southeast Asian dishes, traditional western food and Chinese food, an excellent Chef who specializes in Chinese and western food.

材料 各种蘑菇300克，洋葱30克，蒜5克，黄油30克，白葡萄酒20毫升，盐、胡椒粉、淡奶油各少许，鸡汤底适量。

制作方法

1. 将各种蘑菇清洗干净，切片；洋葱切丝备用。
2. 锅中加入黄油烧热，下入洋葱丝炒出香味。
3. 加入蘑菇片炒香，再加入白葡萄酒。
4. 加入鸡汤底煮20分钟，再用打碎机搅拌成糊状。
5. 将打好的汤二次加热，再加入盐、胡椒粉、淡奶油调味即可。

Ingredients

All kinds of mushrooms	300 g
Onion	30 g
Garlic	5 g
Butter	30 g
White wine	20 ml
Salt	A little
Pepper	A little
Unsalted butter	A little
Chicken soup	Appropriate

Procedures

1. Clean up all kinds of mushrooms, slice up them. Cut onion into shreds.
2. Put the butter into the pot and heat up. Put onion shreds in it and fry fragrance.
3. Put mushrooms in pot and sauté until fragrant, put into white wine.
4. Put chicken soup and boil for 20 min, stir into paste with broken machine.
5. Heat chicken soup twice, and put salt, pepper, unsalted butter to sauce.

奶油紫薯汤 Cream of Purple Potato Soup

材料 紫薯300克，洋葱30克，黄油30克，蜂蜜15毫升，鸡汤底、淡奶油各适量。

制作方法

1. 将紫薯去皮，切片。
2. 洋葱切丝。
3. 锅中加黄油烧热，下入洋葱丝炒出香味，再加入紫薯片。
4. 加入鸡汤底煮至紫薯软烂，用打碎机打碎。
5. 二次加热后加入适量的蜂蜜和淡奶油调味即可。

Ingredients

Purple potato	300 g
Onion	30 g
Butter	30 g
Honey	15 ml
Chicken soup	Appropriate
Unsalted butter	Appropriate

Procedures

1. Peel the purple potatoes and slice it up.
2. Cut onion into shreds.
3. Put butter into the pot and heat up, put onion and stir and sauté until fragrant, then put the purple potato chips in it.
4. Put chicken soup in it until the purple potatoes boiled, break it with beating crusher.
5. Heat up it twice and put right amount honey and unsalted butter to sauce.

张德斌 | 主管　沈阳市盛贸饭店

Zhang Debin | Supervisor　Shenyang Shengmao Hotel

西餐专业委员会会员，从业11年，先后在沈阳、上海、深圳等十余家国际星级酒店任职，辽宁现代服务技术学院特聘教师，擅长意大利菜 、法国菜、德国菜、印度菜等。

Member of Western-style Food Professional Committee, working for 11 years in Shenyang, Shanghai, Shenzhen and other cities for more than 10 international star hotels, distinguished teachers of Liaoning Modern Service Institute of Technology, skilled in Italian cuisine, French cuisine, German cuisine, Indian cuisine and so on.

培根土豆汤 Bacon Potato Soup

马洪彬 | 厨师长　皇朝万鑫酒店
Ma Hongbin | Leader Chef　Royal Wanxin Hotel

1994年从厨至今。
Cooking since 1994.

材料 土豆250克，培根25克，大葱20克，黄油30克，香叶、盐、胡椒粉、淡奶油各少许，鸡汤底适量。

制作方法

1 将土豆去皮，洗净，切块；大葱切段；培根切末。

2 锅中加入黄油烧热，下入培根末炒至出油，再加入大葱炒出香味。

3 加入土豆块、香叶、鸡汤底煮至土豆软烂，再用打碎机搅拌成糊状。

4 二次加热后用盐、胡椒粉、淡奶油调味即可。

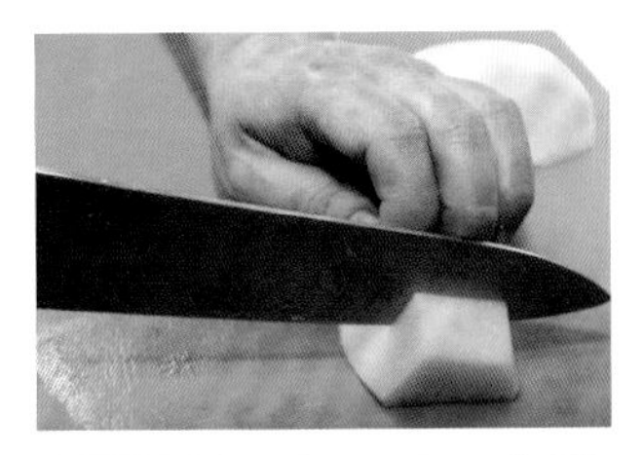

Ingredients

Potato	250 g	Salt	A little
Bacon	25 g	Pepper	A little
Green onion	20 g	Unsalted butter	A little
Butter	30 g	Chicken soup	Appropriate
Myrcia	A little		

Procedures

1 Peel and wash up the potatoes, cut them into slices, cut the green onion into segments, and cut the bacon up.

2 Put butter in the pot and heat up, put the bacon in it and sauté until the oil is out, put the green onion in it and sauté until fragrant.

3 Add the potatoes, myrcia, chicken soup until potatoes are smashed, and then break it with beating crusher.

4 Heat up it twice, add salt, pepper, unsalted butter to sauce.

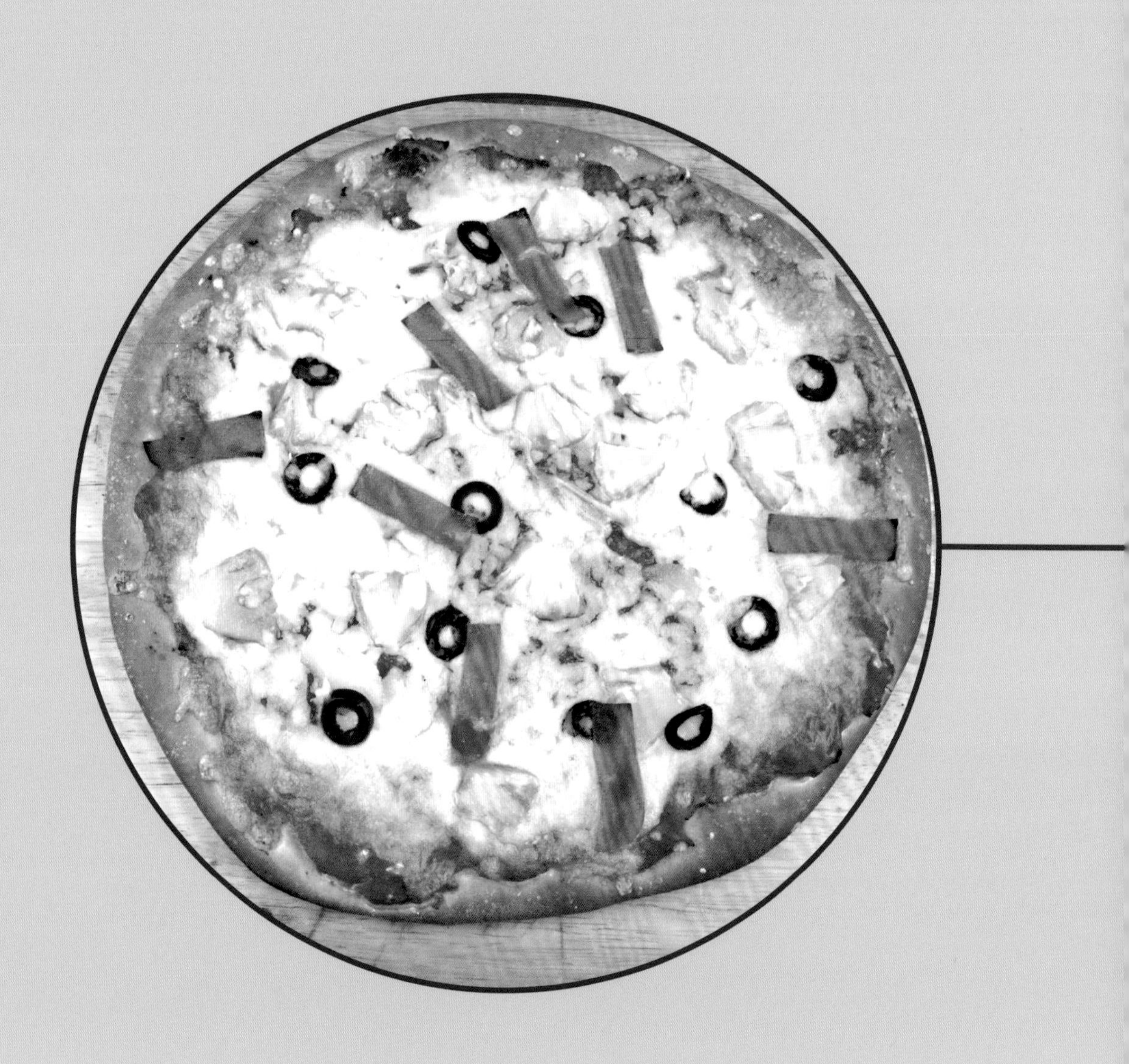

第五章：主食类

Staple Food

肉酱意面 Spaghetti Bolognese

张　旭 | 西餐副厨师长　沈阳丽都索菲特酒店

Zhang Xu | Deputy Chef of Western Food　Shenyang Sofitel Hotel

西餐专业委员会会员，西餐高级烹调师，2004年从事西餐工作至今，曾任职于喜达屋集团、洲际酒店集团和雅高集团，师从马来西亚大师Alexander. Chong。擅长传统法国菜、东南亚菜、意大利菜、德式菜等。

Member of Western-style Food Professional Committee, senior western cook, engaged in western food since 2004, once worked in Starwood, IHG and AccorHotles. Follow Malaysia master Alexander Chong, skilled in Bourgeoise Cuisine, Southeast Asia, Italy, Germany cuisine, etc..

材料 意大利面150克，意大利肉酱60克，意大利番茄酱30克，洋葱末10克，蒜末5克，盐少许，橄榄油适量。

制作方法

1 锅中加水烧沸，放入适量盐，再下入面条煮至八成熟，捞出沥水。

2 锅中加入橄榄油，下入洋葱末、蒜末炒出香味，再加入意大利肉酱和意大利番茄酱烧开。

3 放入煮好的面条，将汁酱收稠，均匀的挂在面条上。

4 加入适量橄榄油拌匀。

5 出锅装盘后搭配帕玛森芝士即可。

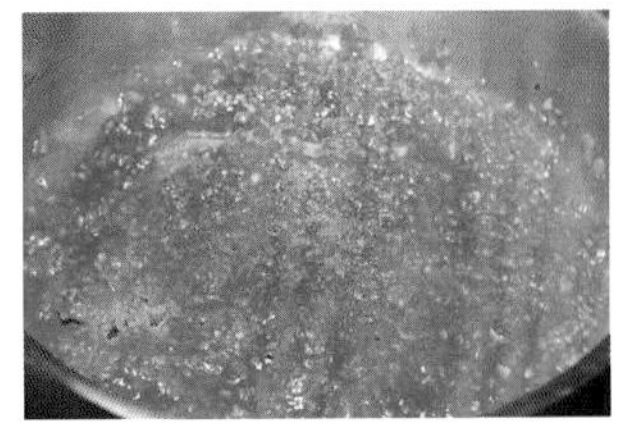

Ingredients

Pasta	150 g	Garlic chops	5 g
Bolognese	60 g	Salt	A little
Ketchup	30 g	Olive oil	Appropriate
Onion chops	10 g		

Procedures

1 Boil the water with appropriate salt and put the pasta to boil 80%.

2 Add olive oil, onion, garlic to sauté until fragrant, put the bolognese and ketchup to boil again.

3 Add the well done pasta, make the bolognese dense, and pour it on the spaghetti.

4 Add olive oil and stir.

5 Place it and add parmesan cheese.

奶汁意面 Cream Spaghetti

材料 意大利面150克，培根30克，洋葱10克，蒜5克，蘑菇20克，鸡蛋黄1个，淡奶油50毫升，盐、黑胡椒碎、法香碎、芝士粉各少许，白葡萄酒15毫升。

Ingredients

Pasta	150 g
Bacon	30 g
Onion	10 g
Garlic	5 g
Mushroom	20 g
Egg yolk	1 pc
Unsalted butter	50 ml
Salt	A little
Black pepper	A little
Parsley flakes	A little
Cheese powder	A little
White wine	15 ml

制作方法

1. 洋葱、蒜均切末；培根切丝；蘑菇切片。
2. 锅中加入橄榄油，下入培根丝炒至出油，再加入蘑菇片、洋葱末、蒜末炒香。
3. 加入盐、白葡萄酒和淡奶油。
4. 加入煮好的意面，将奶油收稠。
5. 离火后加入鸡蛋黄迅速搅拌均匀。
6. 装盘时撒上适量芝士粉、黑胡椒碎和法香碎即可。

Procedures

1. Cut onion and garlic into powder, cut bacon into wire, cut mushrooms into slices.
2. Sauté the bacon with olive oil and add the mushrooms, onion garlic.
3. Add salt, white wine and unsalted butter.
4. Add the pasta and make the cream dense.
5. Add the raw egg yolk and stir quickly after turning off the heat.
6. Sprinkle cheese powder, black pepper and parsley flakes as placing it.

赵　强 | 高级主管　沈阳碧桂园玛丽蒂姆酒店
Zhao Qiang | Senior Executive　Shenyang Maritim Hotel

西餐专业委员会会员，从业10年，先后在沈阳、广东、武汉、大连等地的国际酒店任职。擅长意大利菜、德国菜、东南亚菜，旁通中式菜肴。

Member of Western-style Food Professional Committee, worked with ten years experience,once worked in International Hotel corp. in Guangdong, Wuhan, Dalian etc., skilled in Italian, Germany cuisine and Southeast Asian dishes, as well Chinese food.

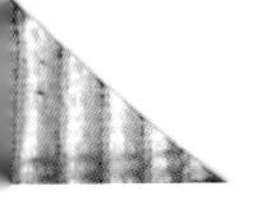

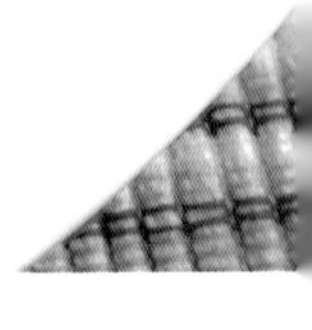

茄汁意面 Spaghetti with Tomato Sauce

材料 意大利面150克，意式番茄酱100克，洋葱末10克，蒜末5克，橄榄油20毫升，盐、糖、法香碎各少许。

制作方法

1 锅中加水烧沸，放入适量盐，下入意大利面煮至八成熟，捞出沥水。

2 锅中加橄榄油，下入洋葱末、蒜末炒香。

3 加入意大利番茄酱烧开，放入盐和糖调味。

4 下入意大利面，让番茄酱均匀地裹在意大利面上。

5 装盘时撒上适量法香碎装饰即可。

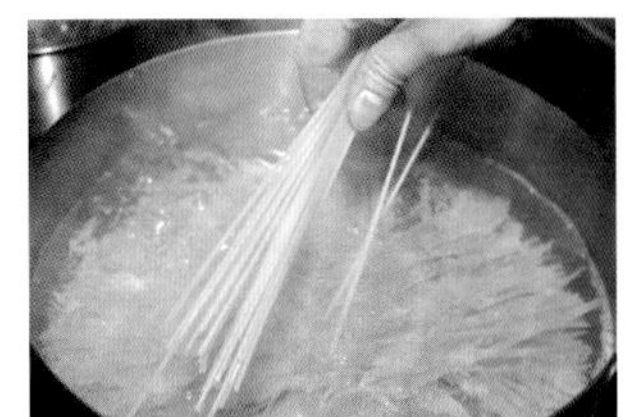

Ingredients

Spaghetti	150 g	Olive oil	20 ml
Italian tomato paste	100 g	Salt	A little
Chopped onion	10 g	Sugar	A little
Chopped garlic	5 g	Thymus vulgaris	A little

Procedures

1 Add water into the pot to boil it, add appropriate amount of salt, put the spaghetti into the pot to boil 80%, remove and drain.

2 Put the olive oil into the pot, put the onion and garlic into the pot, and then sauté until fragrant.

3 Put Italian tomato paste to boil and season with sugar and salt.

4 Put the spaghetti, then make Italian tomato paste smoothed on the spaghetti.

5 Put it in the dish decorated by chopped thymus vulgaris.

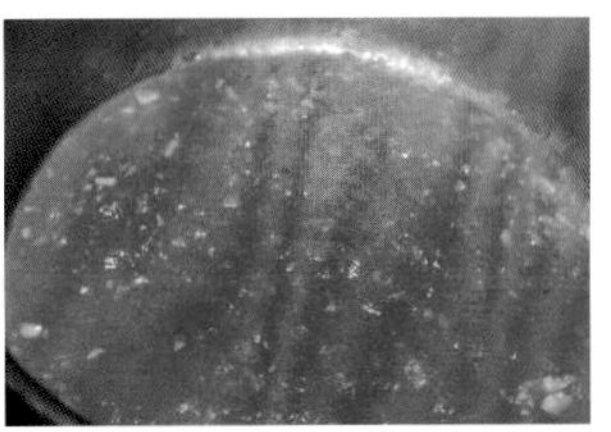

魏上桀 | 厨师长　沈阳碧桂园玛丽蒂姆酒店

Wei Shangjie | Head Chef　Shenyang Maritim Hotel

西餐专业委员会会员，从业16年，先后在日本等地国际酒店任职。擅长日本料理、德国菜、意大利菜。

Member of Western-style Food Professional Committee, worked with 16 years experience, once worked in International Hotel in Japan and etc., skilled in Japanese dish, Germany and Italian cuisine.

夏威夷比萨 Hawaii Pizza

材料 比萨面团220克，火腿、菠萝各30克，橄榄5个，马苏里拉芝士90克，番茄酱50克。

制作方法

1 将火腿、菠萝、橄榄切配好。
2 将面团擀成饼状，用叉子扎上眼。
3 在面饼上抹上一层番茄酱。
4 撒上马苏里拉芝士。
5 芝士上摆放好切配好的原料。
6 放入200℃的烤箱内，将芝士烤制熔化即可。

Ingredients

Pizza dough	220 g
Ham	30 g
Pineapple	30 g
Olive	5 pcs
Mozzarella cheese	90 g
Tomato sauce	50 g

Procedures

1 Chop the ham, pineapple and olive.
2 Roll out the dough into a pie, use the fork to prick it.
3 Put some tomato sauce on the pie.
4 Sprinkle the mozzarella cheese.
5 Put the ingredients chopped on the cheese.
6 Put the pie into the oven, turn the temperature to 200℃, don't take it out until the cheese melts.

刘宗喜 | 厨师长 大石桥凯伦咖啡

Liu Zongxi | Head Chef Dashiqiao Karen Cafe

2002年从事西餐餐饮至今，曾任职于鲅鱼圈区皇家园林、曼哈顿大酒店，师从营口凯伦咖啡实业有限公司行政总厨及技术总指导黄学志老师。擅长法式菜系、意大利菜系。

Engaged in western food since 2002, once worked for Royal Garden Hotel and Manhattan Hotel in Bayuquan Area, studied under the Mr.Huang who is the executive chef and technical guidance of Karen coffee Industries Limited company, skilled in French and Italian cuisine.

魔鬼比萨 Devil Pizza

黄学志 | 总厨师长　营口凯伦咖啡

Huang Xuezhi | Leader Chef　Yingkou Karen Cafe

营口凯伦实业有限公司设计研发部负责人，中国西餐烹饪大师，国家级西餐评委。1999年从事西餐工作至今，擅长法餐、传统意大利菜、日本料理。

Responsible for the Design and Research department of Yingkou Karen industrial limited company, master of western food cooking, national western food judges. Engaged in western food since 1999, skilled in French food., Italian and Japan dishes.

材料 比萨面团220克，意大利萨拉米香肠8片，洋葱、青椒各20克，橄榄5个，银鱼柳5只，芝士90克，番茄酱50克。

Ingredients

Pizza dough	220 g
Italy sausage	8 pcs
Onion	20 g
Green pepper	20 g
Olive	5 pcs
Anchovy	5 pcs
Cheese	90 g
Tomato sauce	50 g

制作方法

1. 将意大利萨拉米香肠、洋葱、青椒、橄榄切配好。
2. 将面团擀成饼状，用叉子扎上眼。
3. 在面饼上抹上一层番茄酱。
4. 撒上芝士。
5. 芝士上摆放切配好的原料。
6. 放入200℃的烤箱内，将芝士烤至熔化即可。

Procedures

1. Prepare with Italy sausage, onion, green pepper and olive.
2. Rolling the dough and then pick some holes by fork.
3. Adorn the pizza with tomato sauce evenly.
4. Add cheese on pizza.
5. Put others ingredients on it
6. Waiting for the cheese flowing in the cave that will be heated on 200℃.

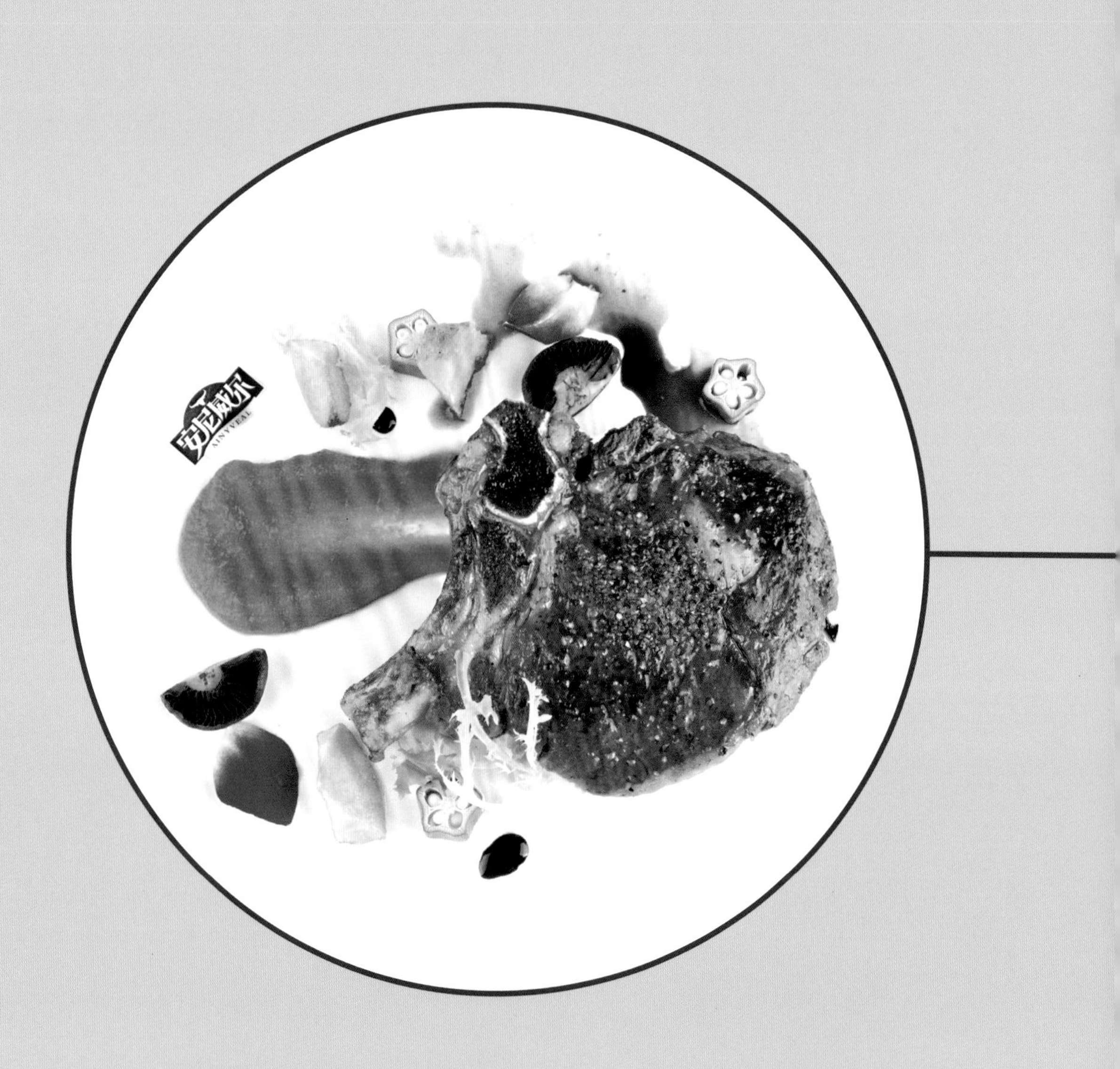
安尼威尔
AINYVEAL

第六章：肉类

Meat

法式烩小牛腱 French Stewed Calf Tendon

邓昌峰 | 首席厨师长 营口凯伦尊尚

Deng Changfeng | Leader Chef Yingkou Karen Noble

西餐专业委员会秘书长，2001年从事餐饮工作至今，擅长中式烹饪料理、传统意大利菜、日本料理。

Leader secretary of Western-style Food Professional Committee, engaged in western food since 2001, skilled in Chinese, Italian and Japan dishes.

材料 小牛腱1个，干葱头10个，口蘑15个，番茄膏20克，迷迭香3克，香叶2片，红酒50毫升，盐、胡椒粉各少许，牛肉汤底适量。

制作方法

1 将小牛腱切块，用盐、胡椒粉、红酒腌制24小时。

2 锅中加油烧热，下入干葱头炒出香味且略微上色，再加入口蘑炒香。

3 加入番茄膏炒2分钟，再加入牛肉汤底。

4 用锡纸覆盖住锅口，放入烤箱，用180℃烤制2小时。

5 装盘前收浓汤汁即可。

Ingredients

Calf tendon	1 pc	Myrica	2 pcs
Dried onion	10 pcs	Red wine	50 ml
Calocybe gambosa	15 pcs	Salt	A little
Tomato paste	20 g	Pepper	A little
Rosemary	3 g	Beef soup	Appropriate

Procedures

1 Cut the calf tendon and pickle it by salt and pepper for 24 hours.

2 Heat the oil in the pot with the dried onion to sauté until being fragnant and colored, and then mix with calocybe gambosa.

3 Add the tomato paste for two minutes, and then add the beef soup.

4 Have the tin foil covering the pan in the oven with 180℃ to bake for two hours.

5 Collect dry soup before dishing up.

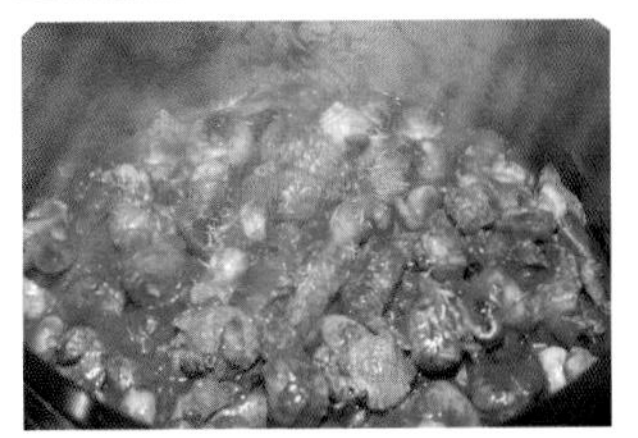

意式小牛腩 Italian Beef Brisket

材料 小牛腩350克，洋葱、西芹、胡萝卜各30克，红酒1/2瓶，橄榄油、盐、胡椒粉各少许。

制作方法

1 将洋葱、西芹、胡萝卜分别切丁。

2 小牛腩用盐、胡椒粉腌制20分钟。

3 将小牛腩放入锅中煎上色。

4 锅中加入橄榄油，放入洋葱丁、西芹丁、胡萝卜丁炒香，再加入小牛腩。

5 加入红酒炖1小时至牛腩软烂即可。

Ingredients

Beef brisket	350 g
Onion	30 g
Celery	30 g
Carrot	30 g
Red wine	1/2 bottle
Olive oil	A little
Salt	A little
Pepper	A little

Procedures

1 Dice the onion, celery and carrot.

2 Pickle the beef brisket with salt, pepper for 20 minutes.

3 Fry the beef brisket in the pot until it colored.

4 Add the olive oil, onion, celery and carrot into the pot to sauté until fragrant, and then add the beef brisket.

5 Add red wine stew for an hour until the flank softens.

柴彦丰 | 西餐厨师　营口凯伦咖啡

Chai Yanfeng | Western-style Food Chef　Yingkou Karen Cafe

2010年从事西餐至今，期间潜心钻研西餐，成为研发部成员，擅长中西融合创意菜、日本料理、传统意大利菜。

Engaged in the western-style food since 2010, during this period studied western-style food and be a member of research and development department, skilled in creative fusion of Chinese and Western food, Japanese cuisine, traditional Italian food.

蔬菜烤小牛展 Vegetables Roasted Calf Exhibition

材料 小牛腱1个，什锦蔬菜200克，盐、黑胡椒各少许。

制作方法

1. 将小牛腱用盐、黑胡椒腌制半小时。
2. 处理好什锦蔬菜。
3. 将什锦蔬菜和小牛腱一起放入200℃烤箱中烤制40分钟。
4. 装盘时淋上牛腱的原汁即可。

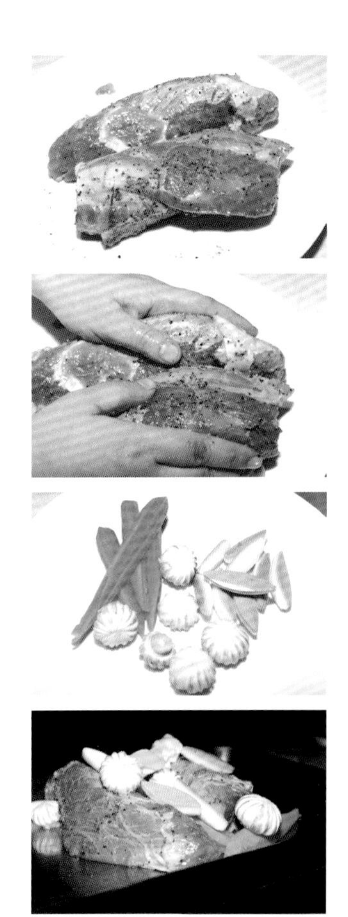

Ingredients

Calf tendon	1 pc	Salt	A little
Assorted mixed vegetables	200 g	Black pepper	A little

Procedures

1. Pickle the calf tendon with salt and black pepper for half an hour.
2. Deal with assorted mixed vegetables.
3. Put the assorted mixed vegetables and calf tendon together into 200℃ oven to roast for 40 minutes.
4. Sprinkle the beef shank soup.

车德蒙 | 厨师长　营口市凯伦咖啡娱乐有限公司总店

Che Demeng | Head Chef　Yingkou Karen Head Office Cafe

2003年加入西餐行业，入职于凯伦咖啡，工作期间一直跟随中国西餐大师、国家级评委黄学志，曾多次学习料理技艺、西餐文化。对各国美食的制作有着浓厚的兴趣。

Engaged in the western-style food since 2003, enter the Karen cafe, followed Chinese western master, natioanl judge Huang Xuezhi during work time, learning cooking skills, western food culture, interested in cooking foreign food.

白汁小牛肉 White Juice Veal

材料 小牛腩300克，牛奶200毫升，鸡蛋3个，面粉、黄油各50克，盐、胡椒粉、香叶各少许。

制作方法

1 将小牛腩切块备用。
2 将小牛腩块放入清水中，加入香叶煮至软烂。
3 锅中加入黄油炒至面粉熟透，再加入牛奶烧开。
4 加入鸡蛋黄迅速搅拌均匀，再下入煮好的小牛腩块。
5 将汁收稠后装盘即可。

Ingredients

Beef brisket	300 g
Milk	200 ml
Egg	3 pcs
Flour	50 g
Butter	50 g
Salt	A little
Pepper	A little
Myrcia	A little

Procedures

1 Cut the beef brisket into pieces and set aside.
2 Put the small pieces of beef brisket in clean water, add myrcia and boil them until softly rotten.
3 Add the butter in pan to fry until the flour is well done, and then add the milk to boil.
4 Add egg yolk quickly stir evenly and then put in the small cooked beef brisket.
5 After the juice is thickened and put it into the plate.

杨文飞 | 厨师长　阜新市凯伦咖啡店
Yang Wenfei | Leader Chef　Fuxin Karen Cafe

西餐专业委员会秘书长，2004年从事餐饮工作，曾任阜新市广宇海鲜皇宫厨房部主管。2013担任阜新市凯伦咖啡店厨师长至今。擅长中式烹饪料理、传统意大利菜、日本料理。

Secretary General of Western-style Food Professional Committee, engaged in food and beverage work in 2004, and served as director of the imperial palace Guangyu seafood kitchen in Fuxin, the director of Karen Cafe so far, skilled in Chinese Cooking Cuisine, traditional Italian dishes, Japanese cuisine.

煎小牛T骨 Fried Calf T-bone

杨 宇 | 餐饮总监 阜新市凯伦咖啡店
Yang Yu | F&B Director Fuxin Karen Cafe

西餐专业委员会秘书长，1999年从事西餐工作。2011年担任阜新市凯伦咖啡店餐饮总监至今，擅长中式烹饪料理、传统意大利菜、日本料理。

Secretary General of Western-style Food Professional Committee, worked in the Western food in 1999, the director of food and beverage at Karen Cafe from 2011, skilled in Chinese Cooking Cuisine, traditional Italian dishes, Japanese cuisine.

材料 小牛T骨1片，土豆片80克，小蘑菇50克，盐、黑胡椒、黑胡椒汁、植物油、黄油各适量。

Ingredients

Calf T-bone	1 pc
Potato chips	80 g
Little mushroom	50 g
Salt	Appropriate
Black pepper	Appropriate
Black pepper sauce	Appropriate
Vegetable oil	Appropriate
Butter	Appropriate

制作方法

1 将小牛T骨用盐、黑胡椒、油腌制10分钟。

2 锅中加油烧热，下入小牛T骨煎至所需程度。

3 将配菜用黄油炒熟。

4 将小牛T骨、配菜装盘，搭配黑胡椒汁即可。

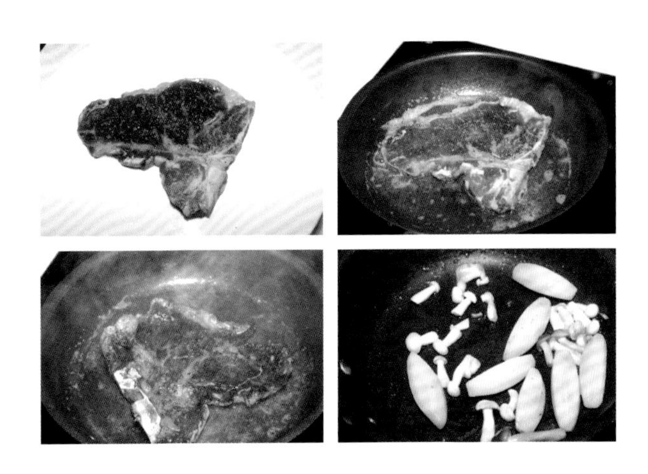

Procedures

1 Calf T-bone is marinated with salt, black pepper, oil for 10 minutes.

2 Put the oil into the pot to heat, fry the calf T-bone to required level.

3 Cook the jardiniere with butter until done.

4 Put the calf T-bone and jardiniere in plates, and match with black pepper sauce.

低温带骨小牛肉眼 Beef Rib Prime of Low Temperature

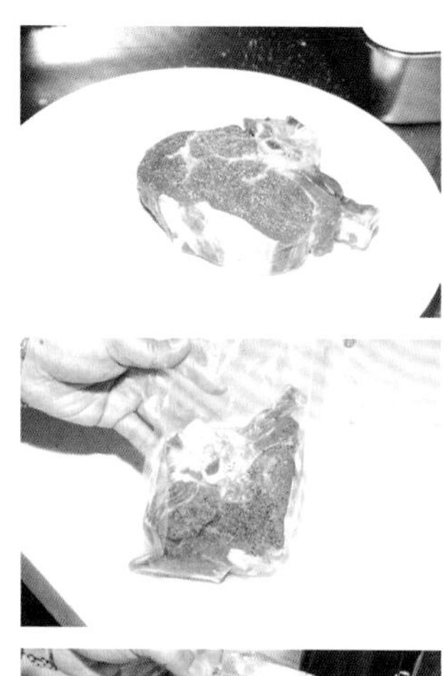

材料 小牛带骨肉眼220克，南瓜泥、土豆、蔬菜各适量，盐、黑胡椒碎各少许，橄榄油20克。

制作方法

1 将牛肉在室温下放置1小时，再撒上盐、黑胡椒碎，腌制10分钟。

2 将牛肉放入食品塑封袋中，加入橄榄油，塑封。

3 将牛肉放入恒温60℃的水中，用低温慢煮1小时。

4 取出牛排，放入油锅中，待两面煎上色，出锅装盘。

5 搭配牛肉原汁，点缀南瓜泥、土豆、蔬菜即可。

Ingredients

Beef rib prime	220 g	Salt	A little
Pumpkin puree	Appropriate	Black peppercorn	A little
Potato	Appropriate	Olive oil	20 g
Vegetable	Appropriate		

Procedures

1 Place the beef at room temperature for an hour, sprinkle with salt and black peppercorn, cure for ten minutes.

2 Place the beef in the food plastic bag, add olive oil, sealed.

3 Put the beef in the water at a constant temperature of 60℃, simmer for 1 hour at low temperature.

4 Take out the steak, put in the pan with oil, color both sides, and dish up.

5 Match with beef juice, intersperse pumpkin pure, potato, vegetable.

王松鹤 | 西餐主管　沈阳国际皇冠假日酒店

Wang Songhe | Western Food Supervisor　Shenyang Crowne Plaza Hotels

西餐专业委员会会员，从业5年，先后在沈阳、杭州等四家国际星级酒店任职。擅长意大利菜、德餐菜系，对日本料理、法国菜、东南亚菜也有研究。

Member of Western-style Food Professional Committee, 5 years of practice in Hangzhou, Shenyang and other four international star hotels, skilled in Italian, German, Japanese, French, Southeast Asian dishes.

芝士牛柳 Cheese Beef Fillet

靳蔚君 | 主厨房主管　沈阳国际皇冠假日酒店
Jin Weijun | Head Chef of Main Kitchen　Shenyang Crowne Plaza Hotels

西餐专业委员会会员，中级烹调师，从业7年，曾就职于多家五星级酒店，对饮食有着独特的爱好，对西式烹饪有着广泛的认识，擅长意大利菜、东南亚菜。

Member of Western-style Food Professional Committee, Intermediate cook, 7 years of practice. Once worked in a number of five-star hotels, have a special interest in food, broad understanding of western style cuisine, skilled in Italian, Southeast Asian dishes.

材料　小牛柳200克，芝士、野生蘑菇各少许，盐、黑胡椒碎各少许，橄榄油20克。

Ingredients

Small beef fillet	200 g
Cheese	A little
Wild mushroom	A little
Salt	A little
Black peppercorn	A little
Olive oil	20 g

制作方法

1. 牛柳在室温下放置1小时，再撒上盐、黑胡椒碎，腌10分钟。
2. 牛柳放入油锅中煎至三成熟，出锅，在两块牛柳中间夹上芝士。
3. 将芝士牛柳放入烤箱中烤制3分钟，取出装盘。
4. 搭配牛肉的原汁和野生蘑菇即可。

Procedures

1. Place the small beef fillet at room temperature for an hour, sprinkle with salt and black peppercorn, Curing for ten minutes.
2. Place the beef fillet in the pan to heat for less ripe, insert cheese in the middle of two beef fillet.
3. Place the beef fillet in the oven to bake for three minutes and dish up.
4. Match with the beef juice and wild mushroom.

烟熏香茅小牛柳 Smoked Beef Citronella

材料 小牛柳120克，柠檬草、香菜各少许，盐、黑胡椒碎各少许，橄榄油20克，老抽10毫升。

制作方法

1. 将牛柳切块，用盐、黑胡椒碎、柠檬草、香菜腌制20分钟。
2. 将腌制好的牛柳用香茅穿上。
3. 将牛肉串放入密封罐里，用苹果木烟熏3分钟。
4. 将牛肉串放入油锅中煎至所需成熟度，再放回密封罐。
5. 上菜前往密封罐中加入烟熏即可。

Ingredients

Small beef fillet	120 g
Lemon grass	A little
Coriander	A little
Salt	A little
Black pepper	A little
Olive oil	20 g
Soy sauce	10 ml

Procedures

1. Cut the small beef fillet into slices, and preserve with salt, black pepper, lemon grass, coriander for 20 minutes.
2. Marinate beef with citronella wear.
3. Place the beef string in the jar, and use the apple wood to smoke for 3 minutes.
4. Beef string will be placed in the pan to fry until desired state, and then put back into the sealed can.
5. Smoke in the sealed can before serving.

吴 艳 | 高级主管 金华万达

Wu Yan | Senior Supervisor Jin Hua Wan Da

多家五星级酒店工作经验，擅长法式菜肴和创意西式菜肴。

More than five star hotel working experience, skilled in French dishes and creative western cuisine.

嫩煎小牛沙朗牛排 Sautéed Veal Sirloin Steak

材料　小牛西冷2片，什锦蔬菜100克，盐、黑胡椒、辣椒粉、植物油各适量。

制作方法

1 将小牛西冷、什锦蔬菜用盐、黑胡椒、辣椒粉、植物油腌制24小时。

2 用肉针把牛排筋络扎透。

3 将什锦蔬菜下入油锅中炒熟。

4 锅中加油烧热，下入牛排煎至需要的成熟度。

5 搭配什锦蔬菜即可。

Ingredients

Grilled beef sirloin	2 pcs	Black pepper	Appropriate
Assorted vegetables	100 g	Chili powder	Appropriate
Salt	Appropriate	Vegetable oil	Appropriate

Procedures

1 Put grilled beef sirloin, assorted vegetables with salt, black pepper, chili powder, vegetable oil and cure for 24 hours.

2 Use a needle to pierce grain meat steak.

3 Put the assorted vegetables into the pan fried.

4 Heat oil in a wok, and fry the steak to maturity.

5 Assorted vegetables can be used as collocations.

张宇涛 | 西餐高级主管　长白山万达假日度假酒店

Zhang Yutao | Senior Western Supervisor　Changbai Mountain Wanda Holiday Hotel

西餐专业委员会会员，多年五星级酒店工作经验，擅长意大利菜、东南亚菜、俄式菜。凭借认真专注的态度，不断地将中西菜式融合创新。

Member of Western-style Food Professional Committee, years of work experience in five star hotels, skilled in Italian, Russian, Southeast Asian dishes. With earnest study attitude, constantly engaged innovative fusion of Chinese and Western dishes.

威灵顿小牛柳 Willington Little Beef

材料 小牛柳200克，帕尔马火腿、果仁各50克，酥皮1张，鸡蛋液150克，洋葱20克，盐、黑胡椒、植物油各适量。

制作方法

1 将牛柳用盐、黑胡椒腌制20分钟。

2 将牛柳下入锅中封煎上色。

3 将帕尔玛火腿刨成薄片；洋葱、果仁切成碎末。

4 锅中加油烧热，下入洋葱碎、果仁碎炒熟，出锅晾凉，放在火腿片上备用。

5 将牛柳用火腿片包裹，再用酥皮包好，放入冰箱内冷藏半小时。

6 将制作好的酥皮牛柳刷上鸡蛋液，在200℃的烤箱内烤制酥皮呈金黄色，取出即可。

Ingredients

Small beef fillet	200 g
Parma ham	50 g
Nutlet	50 g
Pastry	1 pc
Egg liquid	150 g
Onion	20 g
Salt	Appropriate
Black pepper	Appropriate
Vegetable oil	Appropriate

Procedures

1. Marinate small beef fillet with salt and black pepper for 20mins.
2. Put beef into oil pot to fry colored.
3. Cut Parma ham into pieces, and cut off the onion and nutlet.
4. Heat the oil, put Parma ham and onion in pot to sauté, then cool to put on the ham.
5. Roll the beef with nutlet pieces, parcel with pastry and then cool in refregirator for 30mins.
6. Brush egg liquid on it, bake into yellow in 200℃ oven and then take out.

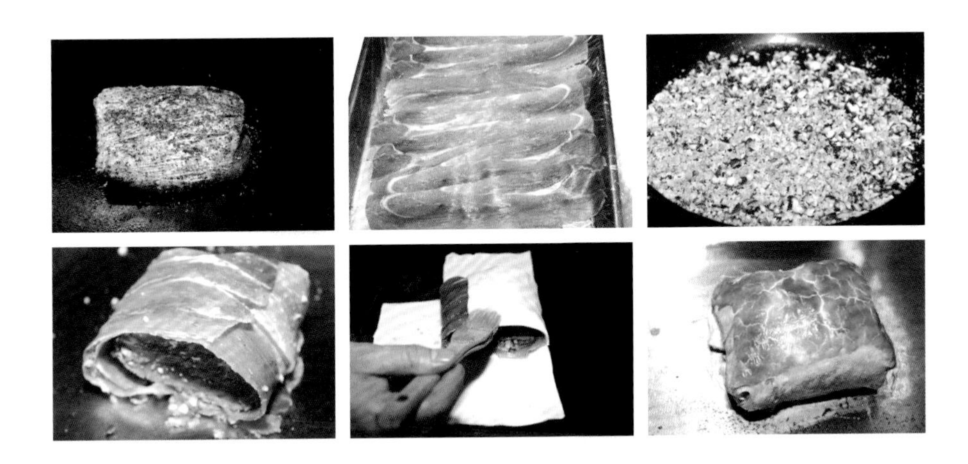

王　刚 | 行政总厨　沈阳阿尔卑斯西餐管理公司
Wang Gang | Executive Chef　Shenyang Alps Western Food Managing Company

拥有20余年西餐工作经验，曾在多家企业酒店任职，擅长法式菜肴。

More than 20 years' working experience, served in many hotels, skilled in French dishes.

意大利烩小牛膝骨 Italy Braised Veal Bone

材料 小牛膝骨2片，洋葱、西芹、胡萝卜各30克，番茄2个，盐、黑胡椒、黄油、面粉各适量。

制作方法

1 将小牛膝骨用盐、黑胡椒腌制10分钟，再拍上一层薄面粉。

2 锅中加入黄油烧热，下入小牛膝骨煎至两面上色。

3 将洋葱、西芹、胡萝卜、番茄用打碎机打成末。

4 将各种蔬菜末放入开水锅中烧沸，再下入煎好的牛膝骨烧至2小时。

5 出锅装盘，搭配意大利烩米饭即可。

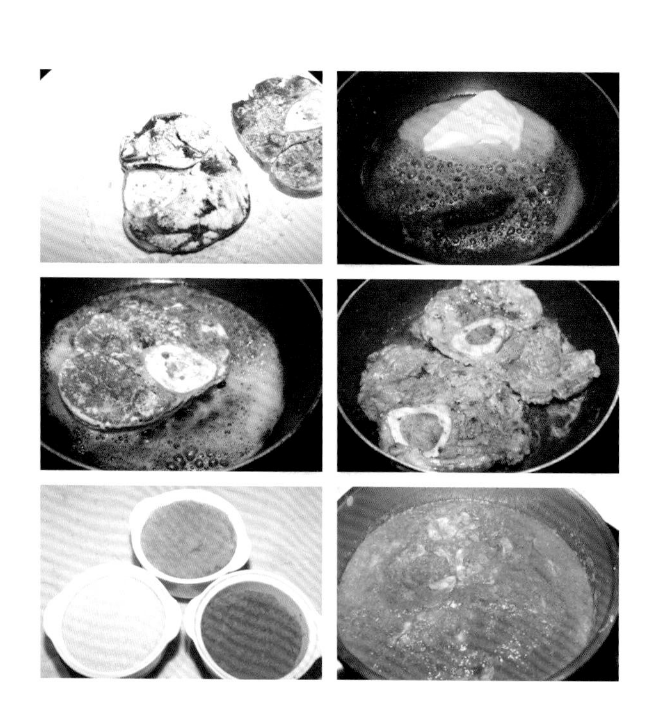

Ingredients

Veal shank	2 pcs
Onion	30 g
Celery	30 g
Carrot	30 g
Tomato	2 pcs
Salt	Appropriate
Black pepper	Appropriate
Butter	Appropriate
Flour	Appropriate

Procedures

1 Cure veal shank with salt, black pepper 10 mins and put thin flour.

2 Pull oil into pot, heat it, put veal shank and then fry it to color it in sides.

3 Cut off onion, celery, carrot, tomato by beater.

4 Put vegetable in heated water pot until boiled; put veal shank in pot for 2 hours.

5 Take it out, dish up, combine with risotto.

贾子明 | 高级厨师 沈阳国际皇冠假日酒店

Jia Ziming | Senior Chef Shenyang Crowne Plaza Hotels

西餐专业委员会会员，新生代西式厨师，国家高级烹调师。在继承传统的同时，又有自己的见地与创新，擅长法国菜、意大利菜。

Member of Western-style Food Professional Committee and new era western food chef, national senior cook. Succeed tradition and keep his creative way, especially skilled in Franch cuisins, Italiy cuisins.

鹅肝牛柳 Foie Gras Beef Fillet

张　楠 | 西餐见习厨师长　沈阳华府酒店

Zhang Nan | Western Food Apprentice Chef　Shenyang Huafu Hotel

西餐高级烹调师，2004年从事西餐工作至今，曾任职于喜达屋集团、美国优势集团、洲际集团、万达集团。师从于多位西餐烹调大师，擅长意大利菜、德式菜、法国菜、东南亚菜及分子料理。

Senior western food chef, engaged in western food since 2004, once worked in Starwood, American Advantage Group,InterContinental Hotels Group and Wanda Group. Learned cooking techniques from professional western food chef, skilled in Italian cuisine, German cuisine, French cuisine, Southeast Asian cuisine and molecular gastronomy.

材料 牛柳1个，鹅肝1片，香梨1个，自制玉米饼1张，盐、黑胡椒各少许。

制作方法

1. 将牛柳和鹅肝用盐、黑胡椒腌制10分钟。
2. 香梨去皮，切块，在糖水中煮熟。
3. 将牛柳和鹅肝放入锅中，煎至所需要的程度。
4. 装盘上桌，搭配红酒汁即可。

Ingredients

Beef fillet	1 pc
Foie gras	1 pc
Bergamot pear	1 pc
Corn cake	1 pc
Salt	A little
Black pepper	A little

Procedures

1. Marinate beef fillet and foie gras with salt, black pepper for 10 minute.
2. Peel the bergamot pear, cut it into pieces, and then cook in sugar water.
3. Put beef fillet and foie gras into the pot, fry until the extent to which you need.
4. Served on plates with red wine juice.

美式肉眼牛排 American Ribeye Steak

材料 肉眼牛排1个，西蓝花100克，蒜10克，盐、黑胡椒、黄油、橄榄油各少许。

制作方法

1. 将肉眼牛排用盐、黑胡椒、橄榄油腌制10分钟。
2. 将西蓝花放入加有少许盐的沸水中焯熟。
3. 锅中加入黄油烧热，下入西蓝花炒熟。
4. 将牛排放入锅中煎至所需的程度。
5. 装盘时搭配西蓝花即可。

Ingredients

Ribeye steak	1 pc
Broccoli	100 g
Garlic	10 g
Salt	A little
Black pepper	A little
Butter	A little
Olive oil	A little

Procedures

1. Marinate ribeye steak with salt, black pepper and olive oil for 10 minute.
2. Put the broccoli in the boiling water with a little salt.
3. Put the butter into the pot, heat up, and fry the broccoli.
4. Put steak into the pot, fry until the extent to which you need.
5. Served on plates with broccoli.

艾圣凯 | 西餐见习厨师长 沈阳华府酒店

Ai Shengkai | Western Food Apprentice Chef Shenyang Huafu Hotel

西餐高级烹调师，2005年从事西餐工作至今，曾任职于喜达屋集团、美国优势集团、洲际集团。师从于多位西餐烹调大师，擅长意大利菜、德式菜、法国菜、东南亚菜及分子料理。

Western food senior chef, engaged in western food since 2005, once worked in Starwood, American Advantage Group,InterContinental Hotels Group. Learned cooking techniques from professional western food chef, skilled in Italian, German, French, Southeast Asian dishes and molecular gastronomy.

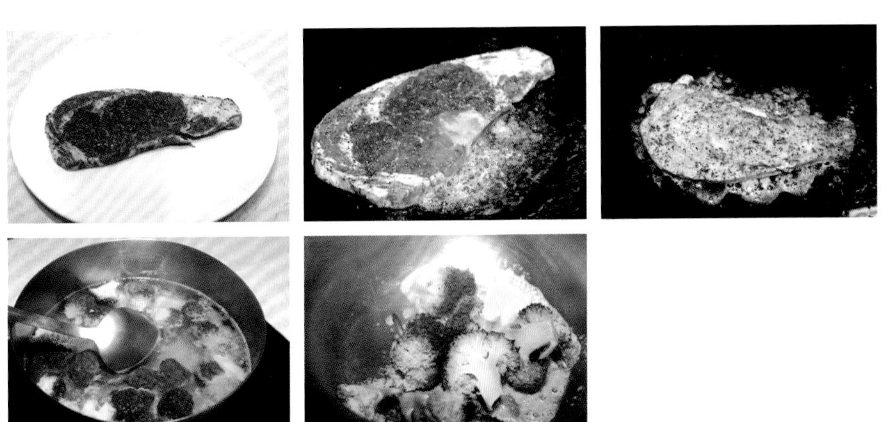

大份T骨牛排 Large T-bone Steak

郑超远 | 主管 沈阳国际皇冠假日酒店

Zheng Chaoyuan | Director Shenyang Crowne Plaza Hotels

从业12年，擅长意大利菜，旁通西式面点。

Practice for 12 years, skilled in Italian dish as well as western style past.

材料 T骨牛排1个，时令蔬菜100克，黑椒汁少许，盐、黑胡椒、黄油、橄榄油各适量。

制作方法

1 将T骨牛排用盐、黑胡椒、橄榄油腌制10分钟。

2 将时令蔬菜切配好，用黄油炒熟。

3 锅中加入黄油烧热，下入T骨牛排煎至所需的程度。

4 出锅装盘，搭配炒熟的时令蔬菜和黑椒汁即可。

Ingredients

T-bone steak	1 pc	Black pepper	Appropriate
Seasonal vegetables	100 g	Butter	Appropriate
Black pepper sauce	A little	Olive oil	Appropriate
Salt	Appropriate		

Procedures

1 Pickle T-bone steak with salt, black pepper, olive oil for 10 minutes.

2 Cut the seasonal vegetables and sauté with butter.

3 Put the butter into the pan and heat it up. Then put in the T-bone steak and fry until the required degree.

4 Dished up the steak with cooked seasonal vegetables and black pepper sauce.

红酒烩牛肉 Braised Beef with Red Wine

材料 牛柳300克，洋葱块、西芹丁、胡萝卜丁各30克，蒜10克，红酒50毫升，番茄膏25克，盐、黑胡椒、植物油各少许，牛汤底适量。

Ingredients

Beef fillet	300 g
Diced onion	30 g
Diced celery	30 g
Diced carrot	30 g
Garlic	10 g
Red wine	50 g
Tomato paste	25 g
Salt	A little
Black pepper	A little
Vegetable oil	A little
Beef stock	Appropriate

制作方法

1 将牛柳切块，放入碗中，加入洋葱块、西芹丁、胡萝卜丁，再倒入红酒拌匀，腌24小时。

2 将腌好的牛肉块放入锅中煎上色。

3 锅中加油烧热，下入洋葱块、西芹丁、胡萝卜丁炒出香味。

4 加入牛肉块、番茄膏炒3分钟，再加入牛汤底。

5 烩至1小时，收稠汤汁调味即可。

Procedures

1 Dice the beef fillet and put in a bowl, then put in the diced onion, celery and carrot. Pour red wine and mix thoroughly, then pickle for 24 hours.

2 Put in the marinated diced beef and frycolored.

3 Heat the oil in the pan and then put diced onion, celery and carrot into the pan, stir until you smell delicious flavor.

4 Put in diced beef, tomato paste and sauté for 3 minutes. Then pour the beef stock.

5 Simmer for one hour, thicken the stock and flavor.

王伟诗 | 节目嘉宾主持人　辽宁广播电视台

Wang Weishi | Guest Host　Liaoning Radio and Television Station

西餐专业委员会名誉副会长，国家高级烹调师、营养师、药膳师，法国蓝带国际烹饪大师，从事厨师工作16年。

The Honorary Vice President of Western-style Food Professional Committee, national Senior Cook, national senior dietitian, a medicined diet cook, Le Cordon Bleu international culinary master, work as a cook for 16 years.

香肠拼盘 Sausage Platter

材料 各种德国香肠适量，土豆泥、酸椰菜各100克，牛奶25毫升，黄油15克，盐、胡椒粉各少许。

制作方法

1 将香肠在开水中煮3分钟。

2 将控干水分的香肠用黄油煎上色。

3 将牛奶烧开，加入黄油、盐、胡椒粉、土豆泥搅拌均匀。

4 酸椰菜放入锅中炒热备用。

5 装盘，搭配法式芥末即可。

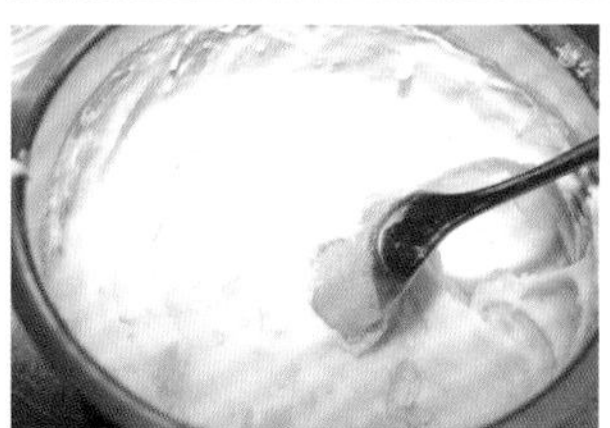

Ingredients

Different kinds of German sausages	Appropriate	Milk	25 ml
		Butter	15 g
Mashed potato	100 g	Salt	A little
Sauerkraut	100 g	Pepper	A little

Procedures

1 Boil the sausages in the boiling water for 3 minutes.

2 Use the butter to fry the dried sausages colored.

3 Boil the milk, add butter, mashed potato, salt and pepper and then stir well.

4 Cook the sauerkraut for spare.

5 Dish up and go with French mustard.

冯玉杰

Feng Yujie

1996年开始曾就职于沈阳商贸饭店、浙江义乌锦都酒店、丽都喜来登酒店、凯宾斯基酒店、抚顺万达嘉华酒店等，对各种餐饮形式如酒店宴会、法式零点、美式快餐、私人订制营养餐配送等有丰富的经验。

Working Since 1996 in Shenyang Trade Hotel, Yiwu Kingdom Hotel, Sheraton Lido Hotel, Kempinski Hotel, Fushun Wanda Hotel etc., experienced in various forms of caterers like hotel banquet, French snacks, American fast food, private customized nutritional meal.

普罗旺斯羊排 Provence Lamb

材料 羊排3片，普罗旺斯烩蔬菜100克，面包糠、松仁碎各少许，大藏芥末20克，盐、黑胡椒、橄榄油、法香各少许。

制作方法

1. 将羊排用盐、黑胡椒、橄榄油腌制10分钟。
2. 把腌制好的羊排封煎上色。
3. 将煎好的羊排抹上芥末，粘上混合好的面包糠、松仁碎、法香。
4. 将羊排在200℃的烤箱中烤制所需的程度。
5. 装盘配上羊肉汁即可。

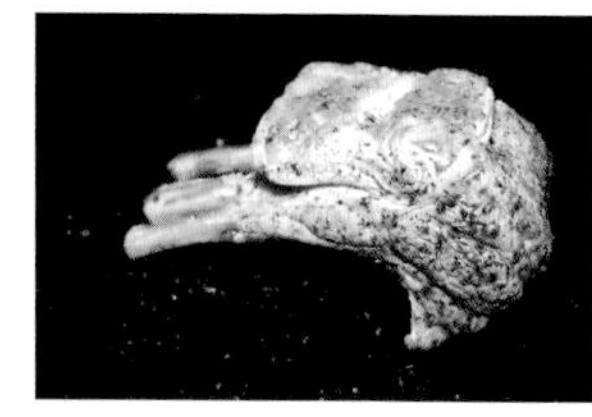

Ingredients

Lamb chops	3 pcs
Provence braised vegetables	100 g
Breadcrumbs	A little
Pine nut	A little
Dijon mustard	20 g
Salt	A little
Black pepper	A little
Olive oil	A little
Parsley	A little

Procedures

1. Marinate the lamb chops in salt, black pepper, olive oil for 10 minutes.
2. Fry and paint the color on marinated lamb chops.
3. Spread the lamb chops on mustard, stick the mixed breadcrumbs with pine nut and parsley.
4. Roast the lamb chops in the 200℃ oven till needed.
5. Dish up with mutton juice.

赵华阳

Zhao Huayang

从业17年，在多家星级酒店和大型连锁餐饮机构任厨师长，擅长法餐、意大利餐、东南亚餐等。

Working for 17 years, leader chef of many star hotels and large chain catering organizations, skilled in French food, Italian food, and Southeast Asian dishes.

低温大虾配柠檬泡沫 Cool King Prawn with Lemon Foam

材料 大明虾1只，柠檬1个，卵磷脂5克，盐、黑胡椒、橄榄油各少许。

制作方法

1. 将大明虾清理干净，用盐、黑胡椒、橄榄油腌制10分钟。
2. 用塑封袋将大虾塑封。
3. 将塑封好的大虾在67℃的水中温煮12分钟。
4. 混合柠檬汁水和卵磷脂，用打碎机打出泡沫。
5. 装盘，搭配时令蔬菜即可。

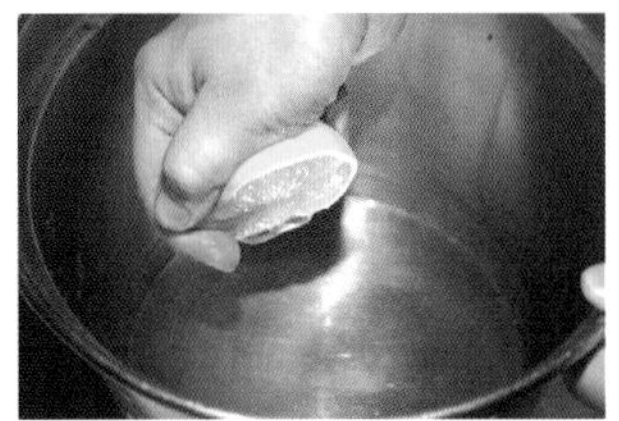

Ingredients

King prawn	1 pc	Salt	A little
Lemon	1 pc	Black pepper	A little
Lecithin	5 g	Olive oil	A little

Procedures

1. Clean the prawn and set it in salt, black pepper and olive oil for 10 minutes.
2. Cover the king prawn with a plastic bag.
3. Poach the prawn in 67℃ water for 12 minutes.
4. Mix the lemon and lecithin and whip it by a juicer.
5. Dish up with some seasonal vegetables collocation.

宋　萌 | 西餐高级主管　景迈柏联

Song Meng | Senior Supervisor　Jing Mai Bo Lian

擅长中西合璧意境菜以及分子料理。

Have a good head of combining Chinese and Western artistic conception dishes and Molecular gastronomy.

法式烤三文鱼腹 French Roasted Salmonbelly

材料 三文鱼腹200克，红椒30克，香菜20克，柠檬半个，盐15克。

制作方法

1. 将三文鱼腹去鳞，清理干净。
2. 用盐将鱼腹的两面抹匀，腌10分钟备用。
3. 红椒洗净，切丁；香菜去梗，洗净，切末，与红椒丁混合，再挤入少许柠檬汁拌匀，装盘。
4. 将腌好的三文鱼腹背面朝上放入烤箱中，用240℃烤10分钟呈金黄色。
5. 将鱼翻过来再烤4分钟，出炉装盘即可。

Ingredients

Salmon-belly	200 g
Cayenne pepper	30 g
Coriander	20 g
Lemon	1/2 pc
Salt	15 g

Procedures

1. Scale the salmon-belly, and clean up.
2. Spread both sides of the salmon-belly with salt and pickle for 10 mins for use.
3. Clean up the cayenne pepper, and cut into small cubes, get rid of the coriander's stems, clean up and chop it, mix with pepper cubes and add some lemon and stir evenly, take into plate.
4. Take the pickled salmon-belly into oven with the back to upside, roast with 240℃ for 10 mins to golden.
5. Turn the salmon-belly and roast for 4 mins, and then take out and take into plate.

郭　清 | 西餐厨师长　南京城市名人酒店

Guo Qing | Leader Chef of Western Food　Celebrity City Hotel

擅长传统法式菜肴，旁通东南亚菜系。

Skilled in traditional French dishes, minor in Southeast Asian dishes.

美式烧烤猪排骨 American Barbecue Pork Ribs

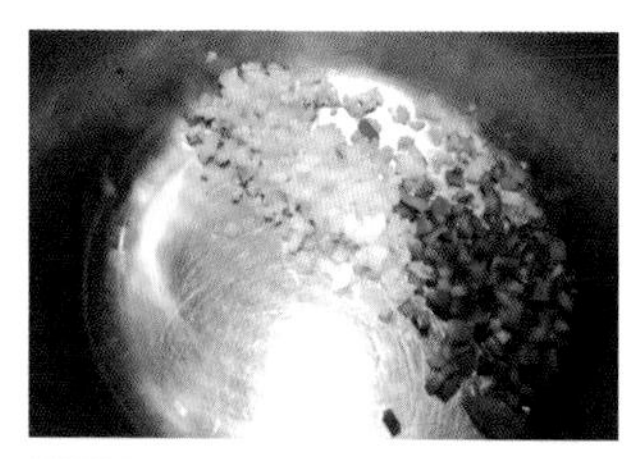

材料 猪排骨250克，土豆块100克，小番茄5个，洋葱末20克，蒜末5克，青尖椒末15克，菠萝30克，BBQ汁20克，番茄沙司15克，盐、黑胡椒、黄油各少许。

制作方法

1. 混合BBQ汁、番茄沙司、盐、黑胡椒、洋葱末、蒜末、青尖椒末，调制成烧烤汁。
2. 将猪排骨放入烧烤汁中腌制24小时。
3. 将腌制过的猪排骨在280℃的烤箱中烤制40分钟。
4. 锅中加入黄油烧热，下入土豆块、小番茄炒熟。
5. 装盘，搭配适量的菠萝即可。

Ingredients

Pork ribs	250 g	Ananas	30 g
Patatas bravas	100 g	BBQ juice	20 g
Tomato cherry	5 pcs	Tomato sauce	15 g
Minced onion	20 g	Salt	A little
Minced garlic	5 g	Black pepper	A little
Green chili	15 g	Butter	A little

Procedures

1. Blend BBQ juice, tomato sauce, salt, black pepper, minced onion, minced garlic, minced green chilin to get barbecue sauce.
2. Put pork ribs in barbecue sauce and pickle about 24 hours.
3. Bake pork ribs for 40 minutes in the oven at 280℃.
4. Heat butter in the pot and sauté patatas bravas, tomato cherry.
5. Dish up, match with suitable ananas.

关 冲 | 厨师长 Back House葡萄酒文化主题西餐厅

Guan Chong | Leader Chef Back House Wine Culture Themed Restaurant

从事西餐15年，曾跟随广州白天鹅厨师长学习多年。

Working on western food for 15 years and followed leader chef of Guangzhou White Swan for years.

德式猪排 German-style Pig Rid

材料 猪外脊200克，煮土豆100克，柠檬1个，水瓜柳20克，鸡蛋液适量，盐、胡椒粉、黄油、面粉、面包糠各适量。

制作方法

1 将猪外脊切片，用盐、胡椒粉腌制10分钟。

2 将腌制好的猪外脊拍上面粉，蘸匀鸡蛋液，再裹上面包糠。

3 将黄油放入锅中烧至熔化。

4 放入猪外脊半煎半炸至熟。

5 装盘，搭配煮熟的土豆、柠檬角和水瓜柳即可。

Ingredients

Pig rid	200 g
Boiled potato	100 g
Lemon	1 pc
Caper	20 g
Egg liquid	Appropriate
Salt	Appropriate
Pepper	Appropriate
Butter	Appropriate
Flour	Appropriate
Breadcrumbs	Appropriate

Procedures

1. Slice pig rid and cure with salt, pepper for 10 minutes.
2. Spread flour on marinated pig rid, stir egg liquid, and then wrap breadcrumbs.
3. Put butter into the pan and melt it.
4. Put pig rid in the pot to fry until half cooked.
5. Dish up, match with boiled potatoes, lemon wedge and caper.

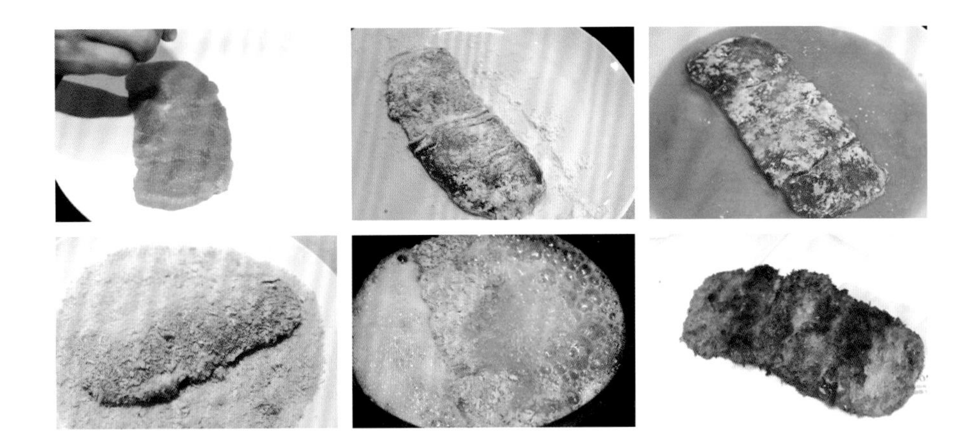

付春贺 | 厨师长　食间牛排

Fu Chunhe | Leader Chef　Shijian Streaking

从事西餐工作13余年，擅长法餐和意大利菜。

Work on western food for more than 13 years and skilled in French food and Italian cuisine.

法式煎春鸡 French Fried Spring Chicken

材料 春鸡1只，洋葱丁、胡萝卜丁、西芹丁各25克，蒜末10克，盐、胡椒粉各少许，甜红粉、五香粉各5克。

制作方法

1. 将春鸡切开一分为二。
2. 混合洋葱丁、西芹丁、胡萝卜丁、蒜末、盐、胡椒粉、甜红粉、五香粉，塞入鸡腹中腌制春鸡，一边腌制一边按摩，揉搓15分钟，让鸡肉更加入味。
3. 将腌制好的春鸡在扒板上煎上色，烤熟即可。

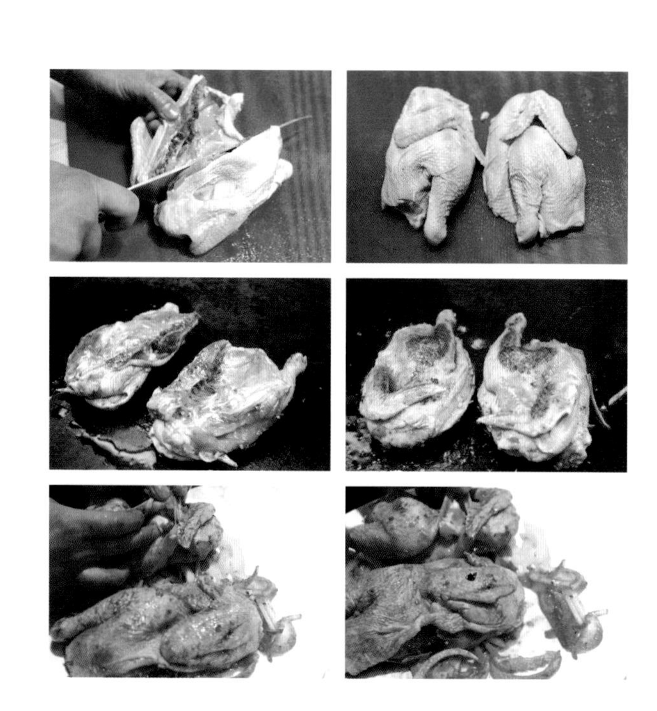

Ingredients

Spring chicken	1 pc
Diced onion	25 g
Diced carrot	25 g
Diced celery	25 g
Minced garlic	10 g
Salt	A little
Pepper	A little
Paprika	5 g
Five spices	5 g

Procedures

1. Cut the spring chicken into half.
2. Mix diced onion, celery, carrot, minced garlic, salt, pepper, paprika, five spices, put the mixture into the chicken's abdomen, salt the chicken, massage it at the same time for 15 minutes to make it tasty.
3. Put the salted chicken on the griddle and fry it to provide the color, until it is done.

刘　强 | 行政总厨

Liu Qiang | Executive Chef

从事西餐美食近20年，曾跟随法国名厨、德国名厨、马来西亚名厨学习西餐菜品制作。在国外学习多年，一直不断地进行菜品创新，不断地追求精致西餐菜品的变革和结合。

Work on western food for almost 20 years, learned with French top chef, Germany top chef, Malaysia top chef for cooking western food. Learning many years for creating dishes and pursuing delicate western dishes reform and combination constantly.

鸡肉农夫三明治 Farmhouse-bread Chicken Sandwich

蔡云龙

Cai Yunlong

从事西餐美食10年，曾在拉斐特城堡酒店担任西餐厨师长，2011年在北京为香港影视剧组做私人厨师，擅长法餐、美餐、意大利餐、艺术创意西餐。

Work on western food almost 10 years and used to be the chef in the Lafayette Hotel, worked as a private chef for Teleplay group of Hongkong in Beijing in 2011, skilled in cooking French, American, Italy cuisine and creative western food.

材料 鸡胸1个，农夫面包1个，生菜25克，黄瓜20克，番茄1个，薯条50克，盐、植物油、卡真粉各适量。

制作方法

1 将鸡胸用盐、植物油、卡真粉腌制五分钟。
2 将鸡肉放入180℃的烤箱中烤至成熟，切成片。
3 农夫面包切片，烤上色。
4 将面包两面抹上蛋黄酱。
5 将生菜、黄瓜、番茄、鸡胸片码放在面包上制成三明治。
6 装盘，配上薯条即可。

Ingredients

Chicken breast	1 pc
Farmhouse bread	1 pc
Lettuce	25 g
Cucumber	20 g
Tomato	1 pc
French fries	50 g
Salt	Appropriate
Vegetable oil	Appropriate
Cajun spice	Appropriate

Procedures

1 Salting chicken breast with salt, vegetable oil, cajun spice for 5 minutes.
2 Put chicken breast in oven which degrees is 180℃ until it is done, and then cut in into pieces.
3 Cut farmhouse bread into pieces, and then toast them to provide color.
4 Mayonnaise the bread on both sides.
5 Put the lettuce, cucumber, tomato, piece of grade breast on the bread to make sandwich.
6 Dishing up with some French fries.

东南亚辣味鸡腿 Southeast Asia Spicy Chicken Leg

材料 去骨鸡腿1个，香米适量，洋葱20克，番茄1个，南姜、泰椒、杏仁片各10克，香茅1根，香菜2克，土豆、胡萝卜各50克，豆蔻1克，盐、鸡精、糖、咖喱粉各少许。

制作方法

1 用盐、咖喱粉、泰椒、南姜、香茅、香菜将鸡腿腌2小时。

2 将鸡腿烤熟，切成块备用。

3 锅中放入洋葱、胡萝卜、杏仁片、豆蔻、香菜、香米，加热焖熟，出锅装盘，再摆上鸡腿块。

4 锅中放入洋葱、番茄、南姜、土豆、香菜、鸡精、咖喱粉煮15分钟，制成咖喱汁，过滤后淋在鸡腿上即可。

Ingredients

Boneless chicken leg	1 pc
Fragrant rice	Appropriate
Onion	20 g
Tomato	1 pc
Galangal	10 g
Thaichili	10 g
Almond slice	10 g
Citronella	1 pc
Caraway	2 g
Potato	50 g
Carrot	50 g
Cardamom	1 g
Salt	A little
Chicken essence	A little
Sugar	A little
Curry powder	A little

Procedures

1 Pickled in salt, curry power, thaichili, galangal, citronella, caraway and cure chicken for two hours.

2 Roast the boneless chicken leg and cut it into pieces for backup.

3 Put the onion, carrot, almond slice, cardamun, caraway and fragrant rice into the pan, heating and braising. Then make them out of the pot on the plate.

4 Put the onion, tomato, galangal, potato araway, chicken essence and curry powder into the pan, and boil them for 15 minutes to be the curry sauce. Finally pour it over chicken leg after filtering.

马 超 | 行政总厨 沈阳领仕扒房
Ma chao | Executive Chef Shenyang Consul Steak House

从业20余年，多次和各国名厨合作，擅长扒制菜肴。

Been a cook for more than 20 years and cooperated with famous cooks all over the world for many times, skilled in grilled dishes.

地中海大虾 The Mediterranean Sea Prawns

孙　鹏 | 西餐总厨　本溪富虹国际饭店
Sun Peng | Leader Chef of Western Food　Benxi Fuhong International Hotel

从事餐饮20余年，曾就职于洲际假日、黎明国际酒店、凯宾斯基饭店等，擅长意餐、德餐、法餐等。

Been engaged in the catering more than 20 years, once worked in InterContinental Hotels Group, Dawn International Hotel, Kempinski Hotel, etc., skilled in Italian food, Geman and French food, etc.

材料 大明虾2只，蒜蓉、小葱末各20克，辣椒末10克，柠檬1个，盐、黑胡椒各少许，白葡萄酒25毫升。

制作方法

1 将大明虾去壳、去虾线，洗涤整理干净。

2 锅中加入黄油烧热，放入蒜蓉和辣椒末炒香。

3 下入大虾煎炒。

4 烹入白葡萄酒，加入盐、黑胡椒调味。

5 撒上小葱末，出锅装盘即可。

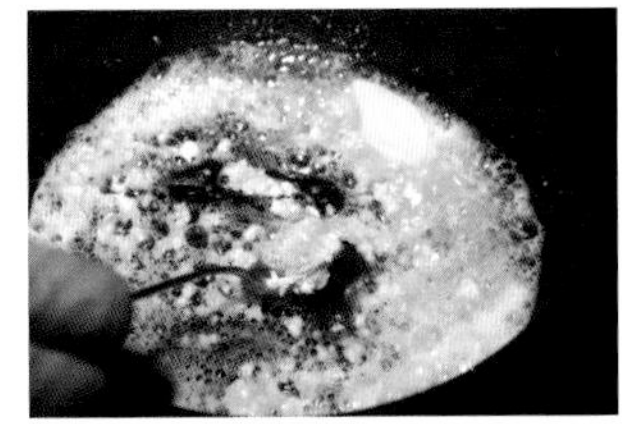

Ingredients

King prawn	2 pcs	Lemon	1 pc
Garlic	20 g	Salt	A little
Chopped green onion	20 g	Black pepper	A little
Pepper chill	10 g	White wine	25 ml

Procedures

1 Remove king prawn shell, shrimp line and wash clean.

2 Add butter to heat with garlic and pepper chill until fragrant.

3 Fry the prawn.

4 Cook with white wine, add salt and pepper to be tasty.

5 Sprinkle with chopped green onion out of the pot on the plate.

烤三文鱼配红花汁 Baked Salmon with Saffron Sauce

材料 三文鱼排200克，土豆块100克，藏红花1克，奶油40毫升，白葡萄酒30毫升，盐、黑胡椒、黄油各少许。

制作方法

1. 将带皮的三文鱼排用盐、黑胡椒腌制10分钟。
2. 将腌制好的三文鱼在扒板上煎至成熟。
3. 锅中加入少许盐和黄油烧沸，放入土豆块煮熟。
4. 混合白葡萄酒、奶油和用开水浸泡过的藏红花制成汁。
5. 装盘，在鱼排上淋上红花汁即可。

Ingredients

Salmon	200 g
Potatoes	100 g
Saffron	1 g
Cream	40 ml
White wine	30 ml
Salt	A little
Black pepper	A little
Butter	A little

Procedures

1. Marinated the skin of the salmon with salt and black pepper for 10 minutes.
2. Marinated grilled salmon in plate and fry to ripe.
3. Add a little salt and butter to the pan, put the potatoes until cooked.
4. Mixed white wine, cream and boiling water soaked in saffron sauce.
5. Installed dish and pour the saffron sauce in fillet.

孟宇飞

Meng Yufei

从事西餐行业13余年，有丰富的酒店工作经验。2008年赴上海深造学习西餐、西点3年，擅长意大利料理。

Engaged in the western food industry for more than 13 years, a wealth of hotel work experience. In 2008, went to Shanghai for further study for 3 years on west food and desserts, skilled in Italy cuisine.

香草黄油煎鳕鱼 Fried Cod with Vanilla Butter

材料　鳕鱼200克，奶油烤土豆100克，黄油15克，法香、蒜各5克，什锦香草2克，盐、黑胡椒各少许。

制作方法

1. 将鳕鱼用盐、黑胡椒腌制10分钟。
2. 将腌制好的鳕鱼放入200℃的扒板上煎至成熟。
3. 用模具将奶油烤土豆卡成圆柱形。
4. 混合黄油、蒜、法香、什锦香草，卷成香草黄油。
5. 土豆放在盘中，上面码放好煎熟的鳕鱼，再放入一片香草黄油即可。

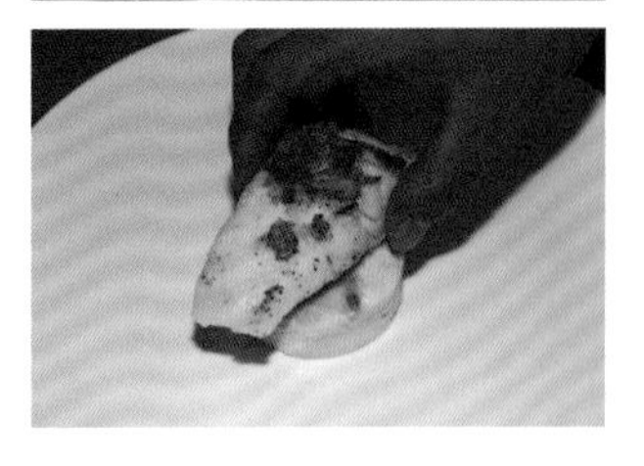

Ingredients

Cod	200 g	Garlic	5 g
Baked potato with cream	100 g	Assorted vanilla	2 g
Butter	15 g	Salt	A little
Fragrance	5 g	Black pepper	A little

Procedures

1. Salt cod with salt and black pepper for 10 minutes.
2. Fry the salted cod in 200℃ and fry until ripe.
3. Put baked potato with cream into a mold.
4. Mix butter, garlic, fragrant, assorted vanilla, to make vanilla butter.
5. Potatoes on the plate with the neatly fried cod, and add vanilla butter.

冯　辉 | 副厨师长　天泊圣汇度假酒店

Feng Hui | Deputy Chef　Tian Bo Sheng Hui Holiday Hotel

2001年毕业于菁华商业管理学院烹饪与营养系专业，曾就职于黎明国际酒店、海韵锦江国际酒店，担任西餐主厨一职。

In 2001 graduated from the School of Business and Administration specializing in the cooking and nutrition, once worked at Dawn International Hotel, Haiyun Jinjiang International Hotel as a western food leader chef.

炸鱼配薯条 Fish and Chips

材料 鲷鱼150克，炸薯条70克，蛋黄酱35克，洋葱末10克，熟鸡蛋碎、柠檬各1个，酸黄瓜碎15克，盐、白胡椒粉、面粉、法香、泡打粉各少许，植物油适量。

制作方法

1. 将鲷鱼切条，用盐、白胡椒粉、法香腌制10分钟。
2. 混合面粉、泡打粉，加适量水制成炸鱼糊。
3. 锅中加油烧至180℃，放入裹匀炸鱼糊的鱼柳炸至成熟。
4. 混合蛋黄酱、洋葱末、熟鸡蛋碎、酸黄瓜碎和适量柠檬汁制成酱汁。
5. 装盘，配上炸薯条和酱汁即可。

Ingredients

Sea bream	150 g
French fries	70 g
Mayonnaise	35 g
Chopped onion	10 g
Cooked egg broken	1 pc
Lemon	1 pc
Sour cucumber	15 g
Salt	A little
White pepper	A little
Wheat flour	A little
Parsley leavers	A little
Baking powder	A little
Vegetable oil	Appropriate

Procedures

1. Cut the sea bream into strips and use salt, white pepper, parsley leaves to pickle for 10 minutes.
2. Mix wheat flour, baking powder, add appropriate amount of water to make fish paste.
3. Add oil to heat until 180℃.
4. Mix mayonnaise, chopped onion, cooked egg, sour cucumber and an amount of lemon juice.
5. Dish up and matched with French fries and sauce.

李东生

Li Dongsheng

从事西餐、调酒行业6年，擅长在传统的西餐中融入现代的元素，宗旨就是不断的创新，引领现代西餐文化。

Engaged western food and wine mixture for 6 years, skilled in the traditional western style food combined with modern elements for constant innovation, leading the modern western food culture.

法式白酒烩贻贝 French White Mussel

材料　青口贝350克，洋葱、黄油各20克，白葡萄酒25毫升，西蓝花80克，法香碎5克，盐、鸡精各少许。

制作方法

1 洋葱切碎；西蓝花掰成小朵。
2 锅中加入黄油烧至温热，放入洋葱碎炒至香软。
3 加入青口贝炒至开口。
4 加入白葡萄酒、盐、鸡精，略煮2分钟。
5 放入西蓝花烧至成熟，即可出锅装盘。

Ingredients

Green mussel	350 g
Onion	20 g
Butter	20 g
White wine	25 ml
Broccoli	80 g
Parsley leavers	5 g
Salt	A little
Chicken essence	A little

Procedures

1 Onion chopped and broccoli breaken into small pieces.
2 Add butter into pot to heat until warm, add onion.
3 Add green mussel.
4 Add white wine, salt, chicken essence to boil 2 minutes.
5 Add broccoli until well-cooked, then dished up.

刘　军 | 厨师长　新民凯伦咖啡
Liu Jun | Leader Chef　Xinmin Karen Cafe

曾就职沈阳飞龙宾馆厨师长、沈阳文化博览园畅馨园宾馆厨师长、沈阳天天海鲜舫主厨、沈阳市吉品特色火锅城厨师长、沈阳华都宾馆厨师长，擅长中西菜品的研发制作，专研牛排披萨的新品研发。

Once worked as leader chef of Shenyang Feilong hotel, Changxinyuan Hotel, Shenyang Tiantian Seafood Restaurant, Shenyang Jipin Special Hot Pot, Shenyang Huadu Hotel, skilled in research and development of Chinese-Western dishes, especially steak pizza.

第七章：日餐类

Japanese Cooking

海鲜炒乌冬面 Sautéed Udon Noodles and Seafood

李　响 | 主厨　新民凯伦咖啡

Li Xiang | Leader Chef　Xinmin Karen Cafe

曾任沈阳华都宾馆主厨、沈阳吉品特色火锅城主厨、沈阳阿兰朵咖啡厅主厨。擅长中西式餐品的制作与研发，喜欢各地的美食文化。

Once worked as a leader chef in Shenyang Huadu Hotel, Jipin Characteristic Hot Pot and Alanduo Cafe. Skilled in producing, researching and developing Chinese and Western style food, be fond of catering culture all over the world.

材料 日式乌冬面1袋，甘蓝、青椒、胡萝卜、香菇各适量，木鱼花15克，植物油15毫升，浓口酱油10毫升，味粉、味淋各5毫升，盐适量。

制作方法

1. 将乌冬面放入开水锅内煮熟，捞出过凉备用。
2. 甘蓝、青椒、胡萝卜、香菇均切成丝备用。
3. 锅中加油烧热，放入各种丝料炒至半熟，再放入乌冬面炒匀。
4. 加入调料炒匀，装盘，上面放上木鱼花即可。

Ingredients

Bag of Japanese udon noodles	1 pc	Bonito flake	15 g
		Vegetable oil	15 ml
Cabbage	Appropriate	Concentrated soy sauce	10 ml
Green pepper	Appropriate	Flavor	5 ml
Carrot	Appropriate	Mirin	5 ml
Lentinus edodes	Appropriate	Salt	Appropriate

Procedures

1. Put the udon noodles in the pot that is filled with boiling water to make the noodles the noodles cooked, and then let it cool.
2. Cut up cabbages, green peppers, carrots, lentims edods in silk.
3. Heat the vegetable oil in the pot, put the silks in it and sauté to hard-boiled, and then place udon noodles and sauté them well.
4. Add the flavoring and sauté them well, put the dish in plate, and then put the bonito flakes on it.

牛肉盖饭 Beef Donburi

材料 肥牛片200克，米饭250克，洋葱70克，香葱末10克，植物油少许，日本照烧汁40毫升，汤底适量。

制作方法

1 洋葱切丝；米饭放入碗内备用。

2 锅中加油烧热，下入洋葱丝炒熟，再加入照烧汁及一勺汤底。

3 放入肥牛片煮熟，盖在饭上，将汤汁浇在上面，撒上少许香葱末即可。

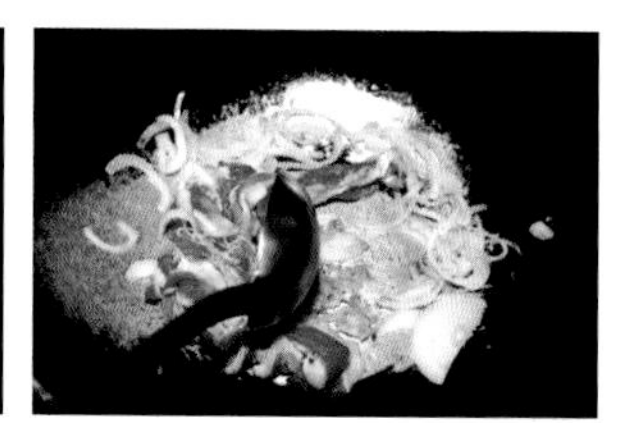

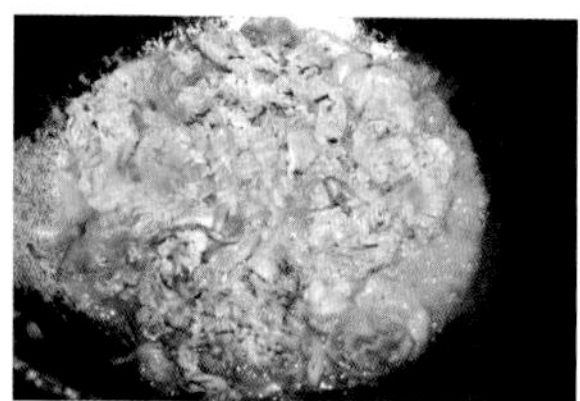

Ingredients

Beef slices	200 g
Rice	250 g
Onion	70 g
Diced chive	10 g
Vegetable oil	A little
Japanese teriyaki juice	40 ml
Soup	Appropriate

Procedures

1 Shred onion and place rice in a bowl for spare.

2 Heat oil in a wok and water into onion silk to stir and sauté, add teriyaki juice and a spoon of soup base.

3 Put into beef slices to cook, cover the rice, pour soup on it and sprinkle diced chive a little.

刘 博 | 主厨 新民凯伦咖啡
Liu Bo | Leader Chef Xinmin Karen Cafe

曾任职于君浩大酒店、大观园酒楼、福满楼酒楼、五洲大酒店的主厨，精通中西餐品，研发出了各式菜肴及西餐糕点。

Once worked in Junhao Grand Hotel, Grand View Garden Boite, Fu Man Lou Boite and Continental Grand Hotel, skilled in Chinese and western food and research various dishes and western pastry.

风味酱汤拉面 Flavor Miso Noodle

材料 日式冷冻拉面1块，叉烧肉2片，熟鸡蛋半个，麻笋、香葱、裙带菜各少许，葱油少许，味增酱50克，猪骨汁100毫升。

制作方法

1. 锅内放入500毫升清水烧开，放入拉面煮熟，捞出备用。
2. 另起一锅，放入500毫升清水烧开，再放入调料煮开，倒入碗内。
3. 将拉面放入汤内，将配料摆在上面即可。

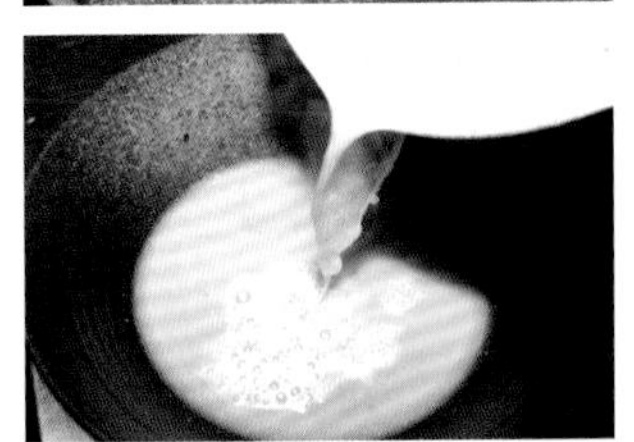

Ingredients

Japanese frozen pulled-noodle	1 pc	Undaria pinnitafida	A little
Barbecue pork	2 pcs	Scallion oil	A little
Cooked egg	1/2 pc	Miso sauce	50 g
Ma bamboo shoots	A little	Pork sauce	100 ml
Chive	A little		

Procedures

1. Bring a pot of water to boil 500ml, put into pulled-noodle to cook, and remove it for spare.
2. Get another pot to boil the water of 500ml and put into flavoring to boil, and then pour into a bowl.
3. Put the pulled-noodle into soup and put the ingredients on it.

段　辉 | 主厨　新民凯伦咖啡

Duan Hui | Leader Chef　Xinmin Karen Cafe

曾任沈阳丽景宾馆、沈阳瑞士海蒂西餐厅主厨，擅长中西式餐品的制作与研发，酷爱西式菜肴的料理以及甜点、各式汉堡、披萨的制作。

Once worked in Shenyang Lijing Hotel and Shenyang Switzerland Heidi Catering, skilled in making and researching Chinese and western dishes, and making western cuisine, dessert, various hamburger and pizza.

醋渍海参 Trepang in Vinegar

材料 活海参1只，香葱少许，日式土佐醋100毫升。

制作方法

1 活海参去内脏，用流水洗干净。

2 锅内加水烧开，放入海参煮15秒，捞出过凉。

3 将海参顶刀切片，放入土佐醋泡20分钟。

4 将海参捞出，放入器皿中，倒入新的土佐醋，没过海参即可。

Ingredients

Live trepang	1 pc
Chive	A little
Japanese indigenous vinegar	100 ml

Procedures

1 Get out entrails of live trepang and wash with running water.

2 Bring to the boil in saucepan, place the trepang in and simmer for 15 seconds, and then get it out and cool by water.

3 Slice up and dip it in indigenous vinegar for 20 minutes.

4 Get it out and place all in plate, pour the vinegar to submerge.

郭　航 | 日厨房厨师长　沈阳国际皇冠假日酒店

Guo Hang | Leader Chef of Japanese Kitchen　Shenyang Crowne Plaza Hotels

西餐专业委员会会员，1988年开始在沈阳鹿鸣春饭店学习中餐，后在日本工作10余年，深谙日本料理的奥义。

Member of Western-style Food Professional Committee and learn to cook in Shenyang Lumingchun restaurant since 1988, and then worked in Japan for 10 years, familiar with Japanese cooking.

日式炸天妇罗大虾 Japanese-fry Tempura Prawn

蒙丹阳 | 日厨房主管　沈阳国际皇冠假日酒店

Meng Danyang | Director of Japanese Kitchen　Shenyang Crowne Plaza Hotels

西餐专业委员会会员，擅长日本料理中的生鱼刺身、铁板烧、煮物及炉端烧的制作。

Member of Western-style Food Professional Committee, skilled in sashimi, teppanyaki, boiling cooking, robatayaki in Japanese cooking.

材料　10头虾5只，紫茄子15克，香菇1个，胡萝卜1片，天妇罗粉300克，天妇罗汁200毫升。

Ingredients

Prawn	5 pcs
Eggplant	15 g
Mushroom	1 pc
Carrot	1 pc
Tempura powder	300 g
Tempura juice	200 ml

制作方法

1 将虾去头、壳，留尾，去虾线，挤断筋备用。

2 茄子切片；香菇切花刀；胡萝卜切片；天妇罗粉用水和匀。

3 将主配料蘸上天妇罗糊，下入热油锅中炸3分钟，捞出沥油。

4 将炸好的天妇罗码入盘中，带天妇罗汁上桌即可。

Procedures

1 Take out heads and shell of the prawn, keep tails, get away bow net and crush tendons up for preparation.

2 Slice eggplant, carrot and mushroom up, stir tempura powder and water well.

3 Dip main ingredient in tempura starch, fry for 3 minutes, and then drip out.

4 Place in plate with Tempura juice.

鸡肉卷 Chicken Rolls

材料 鸡胸肉200克，虾仁100克，蟹足棒、芦笋、紫菜、面包糠、鸡蛋、面粉、泰国鸡酱各适量，盐、味粉、清酒、白胡椒粉各少许。

制作方法

1. 将主料剁成泥，加入调料备用。
2. 将肉泥抹在紫菜上，放入蟹足棒和芦笋，再卷起来。
3. 裹上面粉，蘸匀蛋液，滚面包糠，放入180℃油锅中炸4分钟。
4. 捞出切块摆盘，蘸泰国鸡酱食用即可。

Ingredients

Chicken breast	200 g	Flour	Appropriate
Shrimp	100 g	Thai chicken sauce	Appropriate
Crab foot stick	Appropriate	Salt	A little
Asparagus	Appropriate	Monosodium glutamate	A little
Laver	Appropriate	Sake	A little
Breadcrumbs	Appropriate	White pepper	A little
Egg	Appropriate		

Procedures

1. Chop main ingredient into mud, add flavoring.
2. Wipe muddly flesh on the laver, put in crab foot stick and asparagus, then roll up.
3. Wrap in flour, dip and egg liquid, roll breadcrumbs, and fry with 180℃ oil for 4 minutes.
4. Bail the stripping and slicing and plate, eat with Thai chilli sauce.

藤典军 | 沈阳龙之梦大酒店

Teng Dianjun | Shenyang Longemont Hotel

西餐专业委员会会员，西餐烹调师，2001年至今开始从事西餐烹调工作，曾任职于万豪酒店集团、喜达屋酒店集团。擅长法式菜肴、匈牙利菜肴和东南亚菜肴的制作。

Member of Western-style Food Professional Committee, western food chef, engaged in western food since 2001, once worked in Marriott Hotel and Starwood Hotel, skilled in French, Hungary and Southeast Asian dishes.

芝士焗明虾 Baked Cheese Prawn

材料 6头大明虾2只，芝士、葱头、大蒜各适量，盐、胡椒粉、百里香各少许。

制作方法

1. 大虾除去虾线，背部开刀口，洗净；芝士切碎；葱头、大蒜切粒。
2. 大虾放烤盘中，在背部开口处放盐、胡椒粉、葱头粒、大蒜粒、百里香，再撒上一层芝士碎。
3. 烤箱220℃预热5分钟，将大虾放入烤箱中烤20分钟即可。

Ingredients

Six heads prawn	2 pcs
Cheese	Appropriate
Onion	Appropriate
Garlic	Appropriate
Salt	A little
Pepper	A little
Thyme	A little

Procedures

1. Take out the veins of prawn, open the back, wash and cut up cheese, make onion, garlic into dicing.
2. Put prawn into plate, put salt, pepper, onion, garlic, thyme and sprinkle with a layer of cheese.
3. With 220℃ to heat for 5 minutes as preparation and put the prawn into oven to bake for 20 minutes.

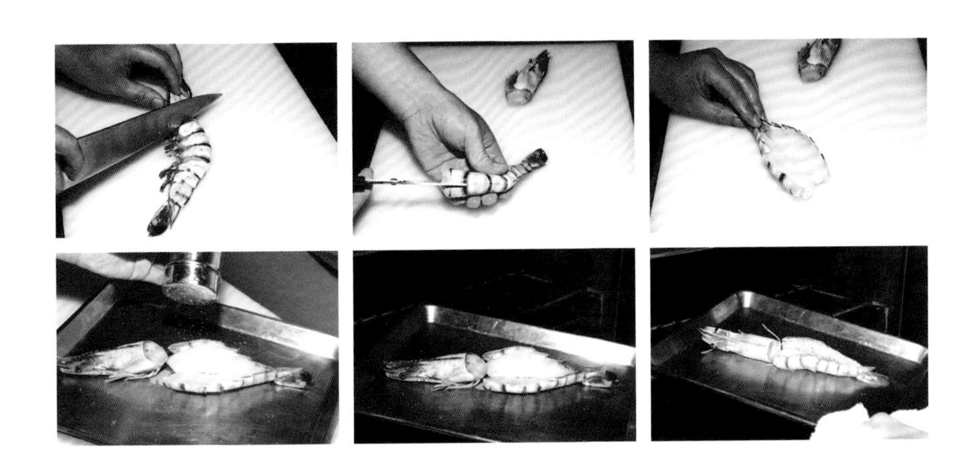

彭　程 | 主管　雀巢公司销售部

Peng Cheng | Director　Sales Department of Nestle

擅长新派意境料理，对于各种食材和调料的搭配有独到的见解。

Skilled in new artistic conception cooking and have the original opinion for the match of ingredient and flavoring.

日式照烧鸡腿饭 Japanese-style Roasted Chicken Rice

材料 去骨鸡腿1个，小沙拉、时令水果、米饭、日式小菜各适量，植物油20毫升，日本照烧汁30毫升。

制作方法

1. 将鸡腿用日式照烧汁腌制20分钟。
2. 将腌制好的鸡腿在200℃的扒板上煎熟。
3. 将日式小菜、沙拉、水果、米饭依次码放在食盒内。
4. 将鸡腿改刀，放入食盒中即可。

Ingredients

Boneless chicken leg	1 pc
Small salad	Appropriate
Seasonal fruits	Appropriate
Rice	Appropriate
Japanese side dishes	Appropriate
Vegetable oil	20 ml
Teriyaki marinade	30 ml

Procedures

1. Pickle the boneless chicken leg with teriyaki marinade for 20 mins.
2. Fry up the marinated chicken on 200℃ grill.
3. Put Japanese side dishes, salad, fruits, and rice in the hamper.
4. Cut the whole piece of chicken leg into small pieces, and put it in the hamper.

余欣佳 | 日厨房主管　沈阳国际皇冠假日酒店

Yu Xinjia | Director of Japanese Kitchen　Shenyang Crowne Plaza Hotels

从业10余年，擅长西式冷菜和日式怀石料理。

Work for more than 10 years and skilled in western cold dish and Japanese kaiseki.

第八章：甜品类

Dessert

芝士蛋糕 Cheese Cake

材料 奶油芝士200克，砂糖75克，鸡蛋液150克，黄油110克。

制作方法

1 芝士和砂糖混合打软。
2 加入鸡蛋液搅匀。
3 加入熔化的黄油搅匀。
4 倒入模具中，隔水160℃烤制，冷却后改刀成长方形，装入盘子中。
5 搭配树莓汁点缀即可。

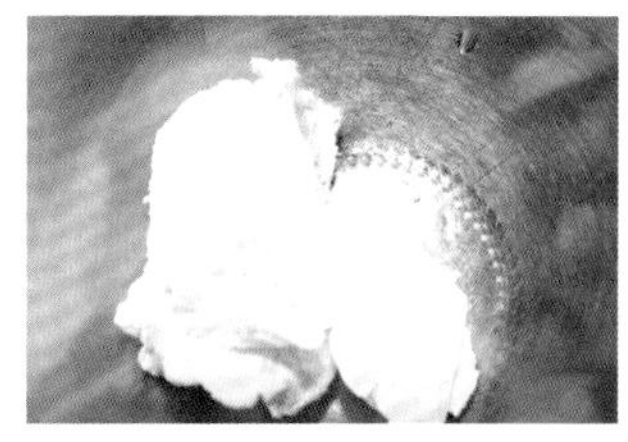

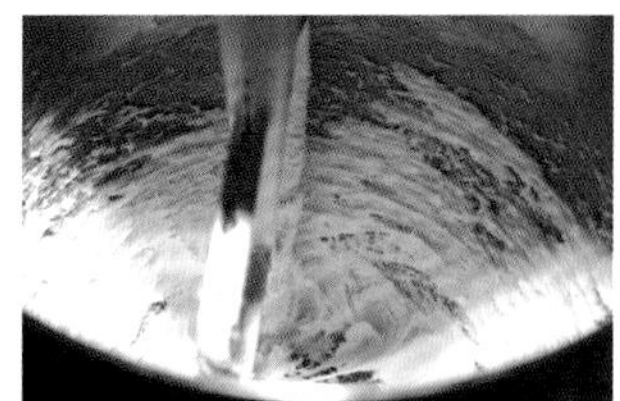

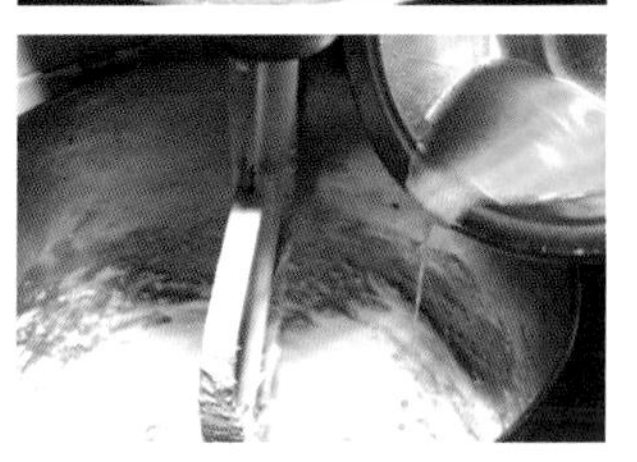

Ingredients

Cream cheese	200 g	Egg mixture	150 g
Sugar	75 g	Butter	110 g

Procedures

1 Mix and soften sugar and cheese.
2 Add in egg mixture and mix it up.
3 Add in melted butter and mix it up.
4 Put it into the mould. Bake at 160℃ separating the water, Cut the shope into rectangle after cooling down and put in the plate.
5 Decorate it with raspberry juice.

韩 笑 | 饼房厨师长 沈阳国际皇冠假日酒店
Han Xiao | Pastry Chef Shenyang Crowne Plaza Hotels

西餐专业委员会会员，从业10余年，一直在国际五星级酒店担任西饼房厨师长，擅长法式甜品和西式面包，对巧克力和糖艺也颇有造诣。

Member of Western-style Food Professional Committee, been in the field for more than 10 years, been the pastry chef in Internatioal Five Star hotel all the time, skilled in French dessert and western-style bread and quite accomplished in making chocolate and sugarart.

马卡龙 Macaroon

材料 糖粉、杏仁粉、糖各300克，蛋清220克，水75克，食用色素少许，奶油适量。

制作方法

1 将糖粉、杏仁粉、110克蛋清拌匀。
2 剩余蛋清打发；糖和水放入锅中煮至121℃，再冲入到打发的蛋清中。
3 将所有原料混合，再放入食用色素混合均匀。
4 用圆形裱花嘴挤在烤垫上，放置4小时至表面成壳。
5 用150℃烘烤20分钟。
6 将制作好的壳中间夹上奶油，摆盘即可。

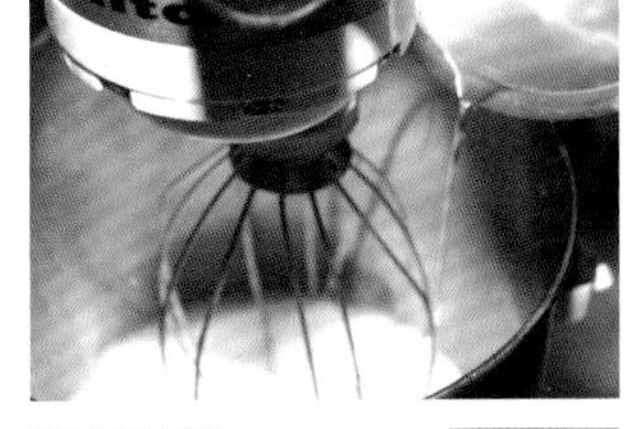

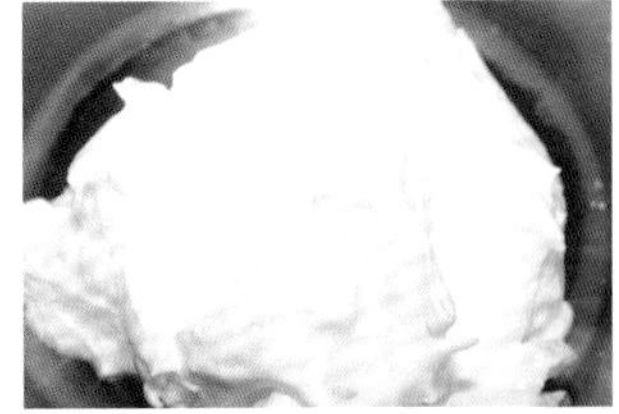

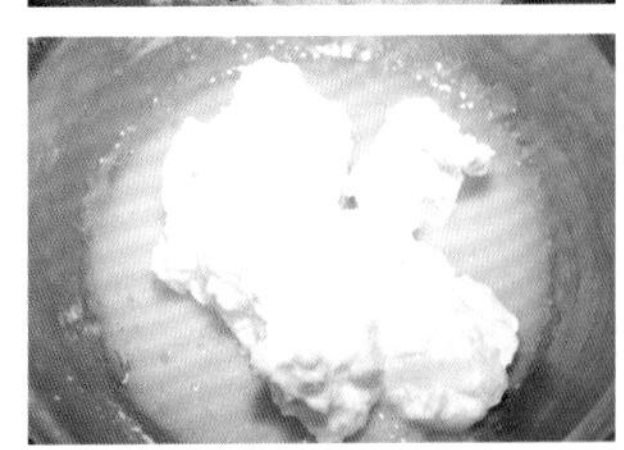

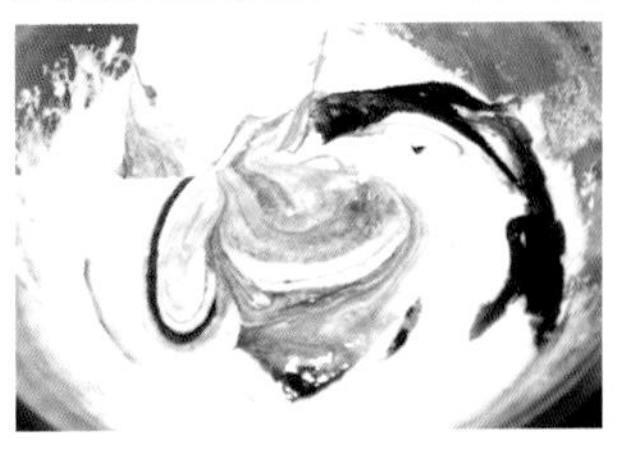

Ingredients

Icing sugar	300 g	Water	75 g
Almond powder	300 g	Edible pigment	A little
Sugar	300 g	Cream	Appropriate
Egg white	220 g		

Procedures

1 Mix up icing sugar, almond powder and 110g egg white.
2 Mix up the rest of egg white. put sugar and water in the pot and cook it to 121℃, pour into the mixed egg whitesd.
3 Mix up all the ingredients, put in edible pigment and mix it up.
4 Squeeze it onto the baking mat with round piping mouth, place it for 4 hours till the surface forms a cover.
5 Bake for 20 minutes at 150℃.
6 Add cream in the middle of the cover and put on dishes.

贾小东 | 饼房主管 沈阳国际皇冠假日酒店
Jia Xiaodong | Pastry Director Shenyang Crowne Plaza Hotels

西餐专业委员会会员，高级烹饪技师，曾获得辽宁省第八届烹饪比赛冠军，辽宁省雕刻第一名，精通各种西点和面包的制作。

Member of Western-style Food Professional Committee, senior cuisine technician, won the champion of The 8th Cooking Competition of Liaoning Province and the first place of sculpture, skilled in making all kinds of western dessert and bread.

巧克力芝士布朗尼 Chocolate Brownie Cheese

材料 黄油70克，奶油芝士190克，糖700克，鸡蛋6个，蛋糕粉250克，杏仁粉180克，可可粉120克，泡打粉、盐各5克，核桃仁200克。

制作方法

1. 容器中放入黄油、奶油芝士、糖混合打发。
2. 逐个磕入鸡蛋。
3. 混合蛋糕粉、杏仁粉、可可粉、泡打粉、盐。
4. 混合步骤2和步骤3中原料，再放入核桃仁。
5. 倒入烤盘中，用180℃烤制成熟，取出放凉，切成块。
6. 装盘，搭配巧克力即可。

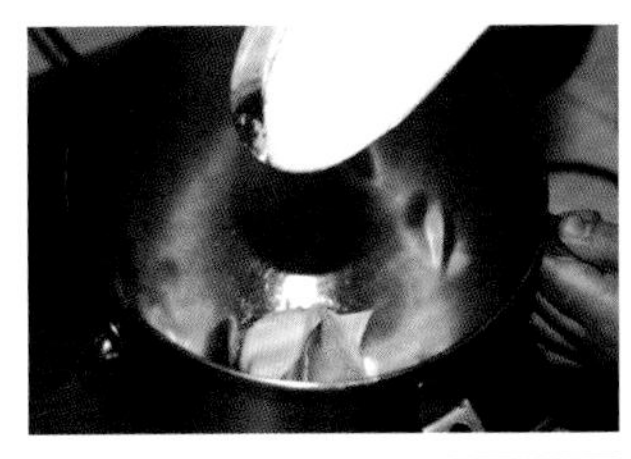
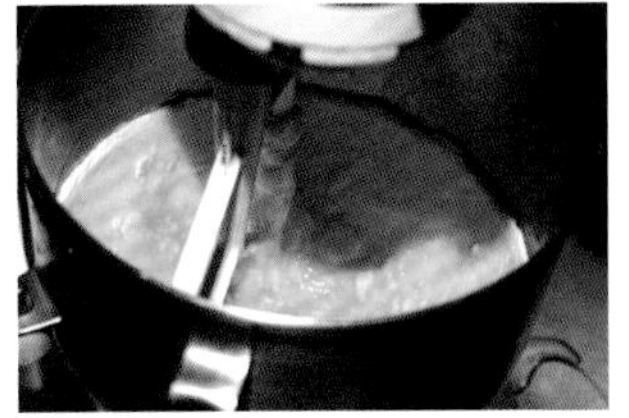
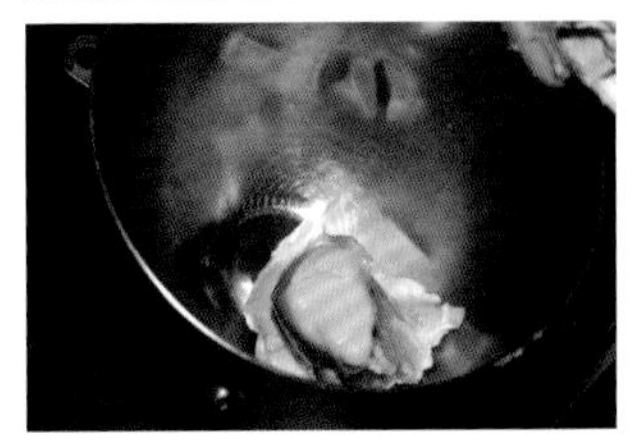

Ingredients

Butter	70 g	Almond flour	180 g
Cream cheese	190 g	Cocoa flour	120 g
Sugar	700 g	Baking flour	5 g
Egg	6 pcs	Salt	5 g
Cake flour	250 g	Walnut meat	200 g

Procedures

1. Put the butter, cream cheese and sugar in container to mix.
2. Put the eggs into container one by one after breaking.
3. Mix cake powder, almond powder, cocoa powder, baking powder and salt.
4. Put the walnut meat after mix materials of step2 and step3.
5. Pour into the pan and bake it with 180℃ to ripe, Cool and cut it into pieces.
6. Serve with collocation of chocolate.

于　月 | 西饼房主管　沈阳国际皇冠假日酒店
Yu Yue | Director of Western Cake Kitchen　Shenyang Crowne Plaza Hotels

西餐专业委员会会员，曾在多家星级酒店任职，擅长法式西点和面包，对甜品的制作和创新有自己的独到见解。

Member of Western-style Food Professional Committee, served in several star hotels, skilled in French cake cooking and have a special understanding of the cooking and innovation of dessert.

拿破仑 Napoleon

材料 面包粉30克，蛋糕粉150克，黄油15克，盐3克，水80克，起酥油100克。

制作方法

1. 混合面包粉、蛋糕粉、黄油、盐和水，打成面团。
2. 用压面机把面团压成长方形面皮。
3. 在面皮中包入起酥油，再压成0.5厘米厚的面皮。
4. 改刀切块，冷冻8小时。
5. 将解冻的酥皮在230℃的烤箱中烤10分钟，取出冷却。
6. 酥皮中加入吉士酱，装盘，搭配新鲜水果即可。

Ingredients

Bread flour	30 g
Cake flour	150 g
Butter	15 g
Salt	3 g
Water	80 g
Shortening	100 g

Procedures

1. Mix the bread flour, cake flour, butter, salt and water into the dough.
2. Press the dough with noodle pressing machine to be the thin and squared.
3. Put the shortening in the bread, and then pressed into crust of 0.5cm.
4. Cut into pieces and freeze for 8 hours.
5. Put the unfrozen crisp surface into 230℃ oven to bake for 10 minutes, and then remove to cool.
6. Add cheese sauce on the crisp surface and serve with fresh fruit.

霍 宁

Huo Ning

从业20年，擅长法式甜品、面包以及糖艺。

Working for 20 years and skilled in French dessert, bread and sugar cooking.

鲜水果塔 Fresh Fruit Tart

张　羽

Zhang Yu

在国际五星级酒店从事饼房工作20年，有着扎实的烘焙专业知识，对蛋糕、面包、巧克力有独到的专业眼光。

Work in bakery kitchen for 20 years in International Five-star hotel, having solid baking professional knowledge and the professional vision at cake, bread, chocolate.

材料 黄油250克，糖125克，鸡蛋5个，水30克，蛋糕粉400克，吉士酱、时令水果各适量。

Ingredients

Butter	250 g
Sugar	125 g
Egg	5 pcs
Water	30 g
Cake flour	400 g
Custard sauce	Appropriate
Seasonal fruits	Appropriate

制作方法

1 将黄油和糖混合打发。

2 逐个加入鸡蛋。

3 倒入水和蛋糕粉制成面团，用压面机压成0.5厘米厚的面皮。

4 将面皮放入塔壳模具内烤制成熟。

5 冷却后在塔壳内挤入吉士酱，用水果装饰即可。

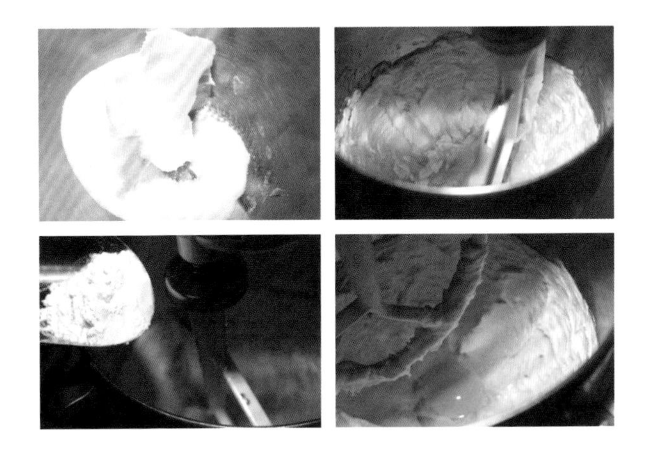

Procedures

1 Mix Butter and sugar together until light and fluffy.

2 Add eggs one by one.

3 Pour water and cake flour to make paste and get the slices that are 0.5 centimeter with the press flour machine.

4 Put the flour slices into mold and bake them.

5 Cool, squeeze the custard sauce into casing mold and decorate it with fruits.

白巧克力慕斯 White Chocolate Mousse

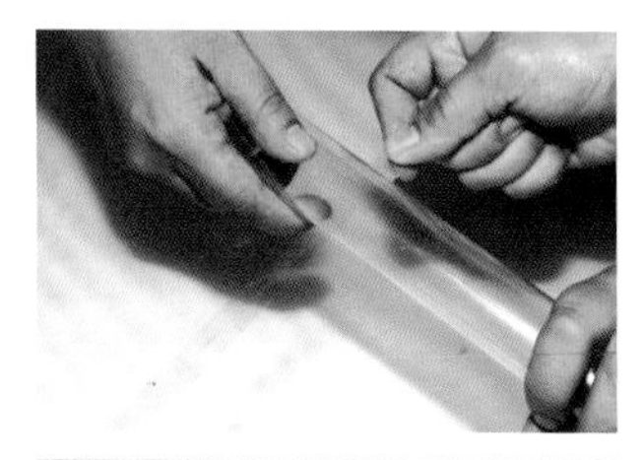
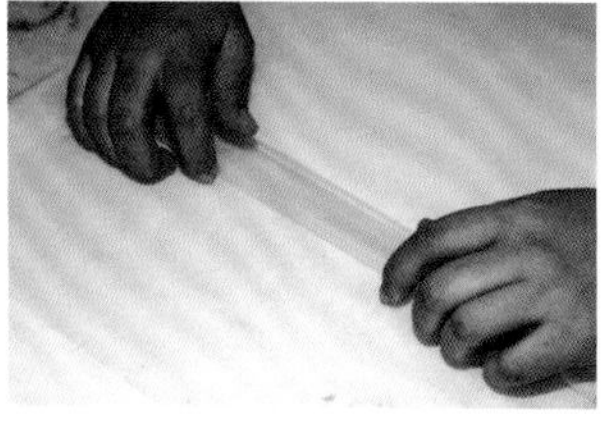

材料 牛奶70克，蛋黄20克，糖15克，白巧克力100克，鱼胶片5克，淡奶油150克。

制作方法

1 牛奶放入锅中煮开，再冲入加有糖的蛋黄中。

2 回火煮85℃，加入熔化的白巧克力。

3 降温到60℃，加入泡好的鱼胶片。

4 降温到40℃，加入打发的淡奶油。

5 倒入模具中，冷冻成型，装盘，搭配果酱和巧克力装饰即可。

Ingredients

Milk	70 g	White chocolate	100 g
Egg yolk	20 g	Gelatin sheets	5 g
Sugar	15 g	Unsalted butter	150 g

Procedures

1 Boil milk with pot and add egg yolk with sugar into it.

2 Boiled to 85℃ and add molten white chocolate.

3 Cool to 60℃ and add gelatin sheets that have been being soaked with water.

4 Cool to 40℃ and add unsalted butter that has been stirred.

5 Pour them into mold, freeze, dish up and serve with jam and chocolate.

曾凡东

Zeng Fandong

在国际酒店管理集团从事饼房工作20年，担任过悦榕庄、万豪、华美达等多家五星级酒店饼房厨师长，有深厚的烘焙专业知识，擅长甜品和巧克力以及面包的创新制作。

Worked in bakery for 20 years in International Hotel Management Group, leader chef of Banyan Tree, Marriott, Ramada and other hotels that are all international five-star hotels, having profound baking professional knowledge and skilled in the creative production of dessert, chocolate and bread.

鲜果香草奶冻 Fresh Vanilla Custard

材料 淡奶油200克，糖40克，香草条1/2条，鱼胶片1片，酸奶100克，橙子1个。

制作方法

1. 将淡奶油、糖放入锅中煮开，再加入香草条浸泡半小时，然后加入鱼胶片；橙子洗净，去皮，切丁。
2. 温度降到30℃，加入酸奶搅拌均匀。
3. 倒入模具中冷冻成型，搭配橙子丁即可。

Ingredients

Unsalted butter	200 g
Sugar	40 g
Herb	1/2 pc
Gelatin sheets	1 pc
Yogurt	100 g
Orange	1 pc

Procedures

1. Put the unsalted butter and sugar and boil, soak the herb for half hour, put the gelatin sheets in. Get the orange washed up, peeled, diced.
2. Put in the yogurt and mix well at 30℃.
3. Put it into the mold until it is frozen and serve with the orange.

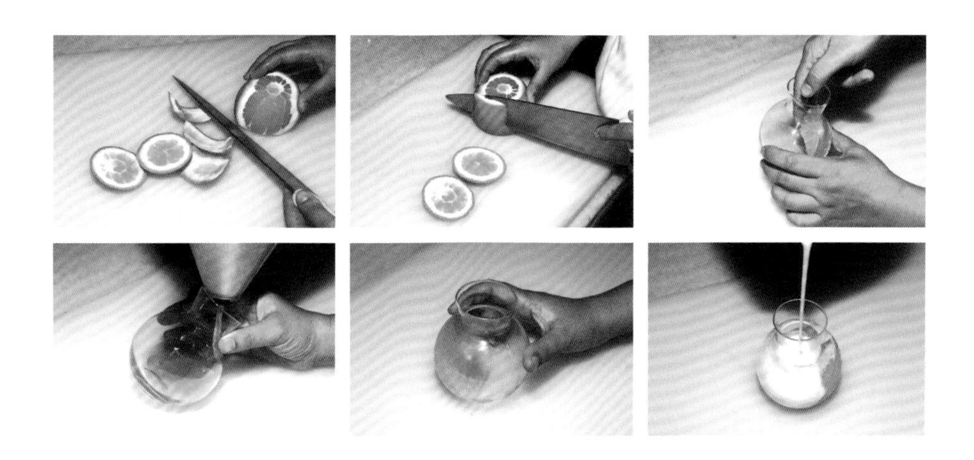

王丽娜 | 逸林饼房厨师长　沈阳希尔顿

Wang Lina | Leader Chef of Yi Lin Pie House　Shenyang Hilton

从事西点技术16年，经验丰富。曾任职于玛丽蒂姆酒店集团、洲际酒店集团、希尔顿集团、喜达屋集团等多家集团酒店。对于甜点追求极致的完美，认真对待每一步制作，为食客的喜悦而喜悦。

Engaged in western bakery for 16 years and experienced. Used to work in Maritim Hotel, Hilton, Starwood Hotel and IHG. Pursue the perfection of dessert cooking and take seriously of every step of the cooking for the joy of customers.

棒子巧克力蛋糕 Hazelnut Chocolate Cake

材料 糖380克，鸡蛋500克，蛋清640克，榛子粉350克，蛋糕粉、榛子酱各100克，榛子碎300克。

制作方法

1 分别打发100克糖、鸡蛋和280克糖、蛋清，打发后混合在一起。
2 加入榛子粉、蛋糕粉、榛子酱、榛子碎搅拌均匀。
3 将混合好的面糊倒入烤盘中，用180℃烤制15分钟。
4 把烤制好的蛋糕坯抹上巧克力酱，切成方形，装盘。
5 搭配鲜奶油和巧克力装饰即可。

Ingredients

Sugar	380 g
Egg	500 g
Egg white	640 g
Hazelnut meal	350 g
Cake flour	100 g
Gianduja hazelnut	100 g
Chopped hazelnut	300 g

Procedures

1 Processed 100g sugar, egg and 280g sugar, egg white and mix.
2 Add the hazelnut meal, cake flour, gianduja hazelnut, chopped hazelnut and mix well.
3 Put the paste into the pan and bake for 15 minutes with 180 ℃.
4 Cover the baked cake embryo with chocolate sauce and diced it and dished up.
5 Serve with cream and chocolate.

文　浩 | 技师　法国安德鲁中国区北区
Wen Hao | Technician　Andrew China North

从事烘焙行业16年，现任法国安德鲁中国区北区技师。曾任职喜达屋、香格里拉、洲际、希尔顿等多家五星级酒店饼房厨师长。精心专研各式甜品，是全能型烘培师。

Work for 6 years and now be the Cooking Technician of Andrew China North. Once served as pastry chef in many five-star hotels such as Starwood Hotel, IHG, Hilton, Shangri-La and so on. Carefully design all kinds of desserts as a well-balanced baker.

焦糖慕斯配鲜草莓 Caramel Mousse with Strawberry

方 圆

Fang Yuan

从业20余年，在多家国际五星酒店担任饼房厨师长。

20 years of professional experience as an executive pastry chef in many five-star hotels.

材料 糖250克，淡奶油950克，鱼胶20克。

制作方法

1. 将糖炒成焦糖，倒入450克淡奶油，用小火煮至融化。
2. 降温至60℃，加入鱼胶片。
3. 降温至40℃，加入打发的500克淡奶油。
4. 将混合的慕斯倒入模具中。
5. 放入冰箱内冷藏8小时，取出装盘。
6. 搭配巧克力和新鲜草莓装饰即可。

Ingredients

Sugar	250 g
Unsalted butter	950 g
Fish gelatin	20 g

Procedures

1. Fry sugar into caramel, pour 450g unsalted butter and boil until melted.
2. Cool down to 60℃ and add fish gelatin.
3. Cool down to 40℃and add 500g unsalted butter.
4. Pour mousse mixture into mold.
5. Put in the fridge for 8 hours and dish it up.
6. Decorate with chocolate and strawberries.

特露芙蛋糕 Trouve Cake

材料 蛋黄200克，蛋清225克，糖200克，盐5克，蛋糕粉100克，可可粉50克，黄油110克，巧克力70克，巧克力奶油适量。

制作方法

1. 分别将蛋黄、100克糖和蛋清、100克糖打发，再混合搅拌均匀。
2. 加入盐、蛋糕粉、可可粉搅匀。
3. 加入融化的黄油、巧克力搅匀。
4. 倒入烤盘中，用180℃烤制12分钟。
5. 将烤制好的蛋糕坯切成三角形，里面夹上巧克力奶油，再用巧克力淋面。
6. 将制作好的蛋糕切成三明治的厚度，装饰巧克力片即可。

Ingredients

Egg yolk	200 g
Egg white	225 g
Sugar	200 g
Salt	5 g
Cake flour	100 g
Cocoa powder	50 g
Butter	110 g
Chocolate	70 g
Chocolate cream	Appropriate

Procedures

1. Whip egg yolk with 100g sugar and whip egg white with 100g sugar, then mix until well incorporated.
2. Add salt，cake flour, cocoa powder and stir until well mixed.
3. Add melted butter , chocolate and mix again.
4. Pour into ovenware and bake for 12 minutes at 180℃.
5. Cut baked cake into triangle, fill with chocolate cream, drizzle with chocolate.
6. Cut cake into thickness of sandwich and use chocolate as decoration.

韩　冬 | 教师　辽宁现代服务技术学院
Han Dong | Instructor　Liaoning Vocational Technical College of Modern Service

西餐专业委员会会员，中西点技师，国家级职业技能鉴定考评员。1999年从事中西面点工作，擅长意式、法式西点制作。

Member of Western-style Food Professional Committee, cook of Chinese and western pastry, judge of national profession test. Working on Chinese and western pastry since 1999 and skilled in Italian and French pastry.

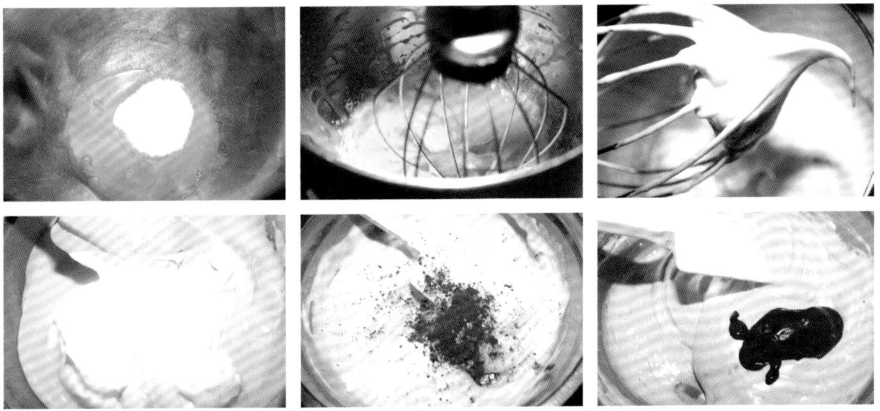

草莓慕斯 Strawberry Mousse

王辰龙 | 技术经理　奥世巧克力公司

Wang Chenlong | Technology Manager　Aalst Chocolate

西餐专业委员会会员，高级西点师。2005年从业至今，曾就职于香格里拉酒店集团、喜达屋酒店集团、洲际酒店集团，擅长法式、意式西点创作。

Member of Western-style Food Professional Committee, senior pastry cook. Engaged in western food since 2005, once worked in Shangri-La Hotel, Starwood Hotel, and InterContinental Hotels Group, skilled in French, Italian pastry creation.

材料 草莓果蓉、牛奶各80克，蛋黄140克，糖70克，鱼胶5克，淡奶油200克。

制作方法

1 将草莓果蓉、牛奶煮开，冲入加有糖的蛋黄中。
2 回火煮到85℃，加入鱼胶。
3 降温到40℃，加入打发的淡奶油，冷却后装入裱花袋。
4 将裱花袋里的草莓慕斯挤到做好的白巧克力球中。
5 搭配新鲜樱桃装饰即可。

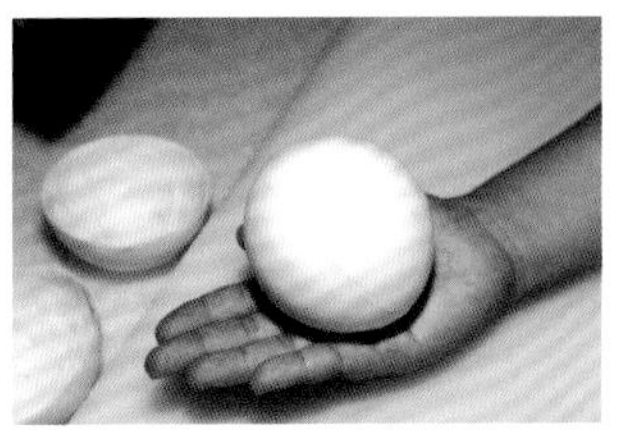

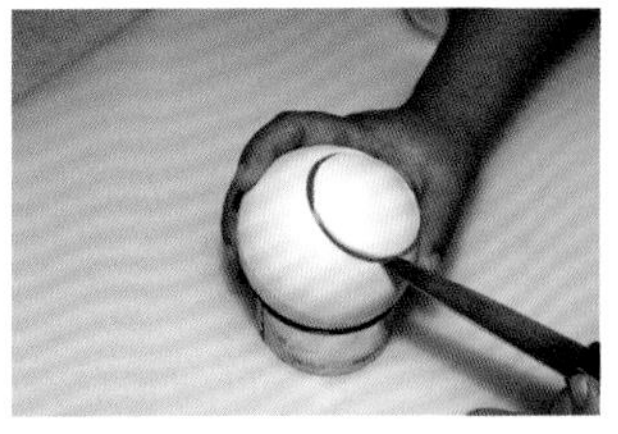

Ingredients

Straswberry puree	80 g	Sugar	70 g
Milk	80 g	Fish gelatin	5 g
Egg yolk	140 g	Unsalted butter	200 g

Procedures

1 Boil the strawberry puree and milk, and put them into egg yolk with sugar.
2 Heat to 85℃ and put into some fish gelatin.
3 Heat to 40℃ and put some beaten unsalted butter in. When cool, put them into pastry bag.
4 Squeeze strawberry mousse in pastry bag to fill the white chocolate ball.
5 Served with fresh cherry.

芒果慕斯 Mango Mousse

材料 糖125克，蛋清90克，芒果蓉300克，鱼胶30克，淡奶油500克。

Ingredients

Sugar	125 g
Egg white	90 g
Mango puree	300 g
Fish gelatin	30 g
Unsalted butter	500 g

制作方法

1. 将糖和少许水放入锅中煮开，再加入打发的蛋清。
2. 芒果蓉中加入打发的淡奶油和鱼胶，混合均匀。
3. 混合步骤1和步骤2中的原料，倒入模具中冷却。
4. 将冷却好的慕斯装盘，搭配杏仁饼干、冰淇淋即可。

Procedures

1. Boil the sugar with a little water and put some egg white foam.
2. Put some unsalted butter foam and fish gelatin into mango puree to mix well.
3. Mix ingredients in step 1 and step 2, and pour into the mold to cool.
4. When cool, serve with almond cookies and ice scream.

史大旺 | 饼房厨师长　沈阳东北大厦

Shi Dawang | Leader Chef of Pastry　Shenyang Northeast Hotel

西餐专业委员会会员，法国蓝带美食协会会员，西式高级面点师。2002年从事饼房工作至今，曾任职于洲际酒店集团、香格里拉酒店集团、索菲特酒店集团。擅长意式、法式、德式甜品面包的制作。

Member of Western-style Food Professional Committee and Le Cordon Bleu Delicacy Association, senior western pastry cook. Engaged in pastry work since 2002, once worked in InterContinental Hotel Group, Shangri-La Hotel, Sofitel Luxury Hotel. Skilled in dessert and bread making of Italian, French and German.

修女泡芙 Nun Cream Puff

材料 水、面粉各250克，黄油200克，鸡蛋8个，吉士奶油适量。

制作方法

1 黄油和水放入锅中煮至沸腾。

2 加入面粉搅拌均匀，离火。

3 面糊降温后分次加入鸡蛋液，搅匀。

4 将面糊装进裱花袋里，挤成圆球状，用200℃烤制30分钟，取出冷却。

5 在冷却的修女泡芙中夹入吉士奶油，装盘，点缀少许巧克力碎片即可。

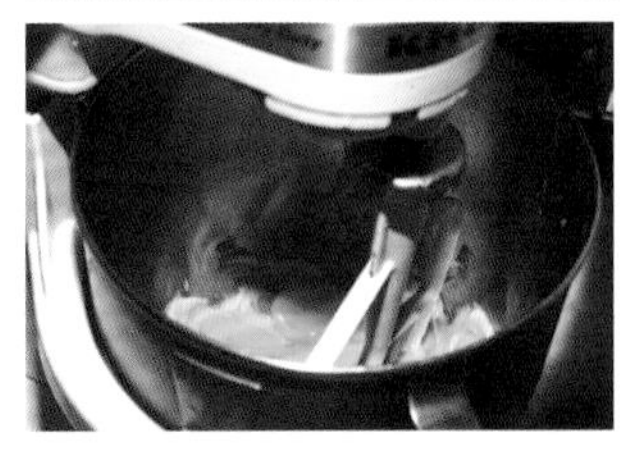

Ingredients

Water	250 g	Egg	8 pcs
Flour	250 g	Custard cream	Appropriate
Butter	200 g		

Procedures

1 Put butter and water into the pot and boil it.

2 Add flour and mix well and be off the fire.

3 Cool the batter, mix with egg liquid, and then stir evenly.

4 Pour the batter into a piping bag, squeeze into ball shape, bake for 30 minutes at 200℃, remove to cool.

5 Add custard cream in cool puff and serve with a little chocolate chip.

张王海 | 饼房厨师长 抚顺万达嘉华酒店
Zhang Wanghai | Pastry Chef Fushun Wanda Royal Hotel

2005年从事西餐饼房至今，曾任职于喜达屋集团、悦榕集团、万达集团。擅长制作传统法式甜品、欧式面包以及巧克力装饰，面艺、糖艺装饰。

Engage in the Western bakery since 2005, once served in the Starwood, Banyan Tree, and Wanda Group. Skilled in making the traditional French dessert, European bread, chocolate decoration, pastry and sugar cooking.

香橙蛋糕 Orange Cake

材料 黄油、糖各250克，鸡蛋6个，蛋糕粉150克，杏仁粉350克，泡打粉5克，橙皮碎少许，浓缩橙汁150克。

制作方法

1 混合黄油和糖，打发，再逐个加入鸡蛋搅打均匀。

2 加入蛋糕粉、杏仁粉、泡打粉、橙皮碎和浓缩橙汁搅匀。

3 将混合好的面糊放入烤盘中，用170℃烤制25分钟。

4 将冷却的蛋糕用模具制成圆形，搭配白巧克力即可。

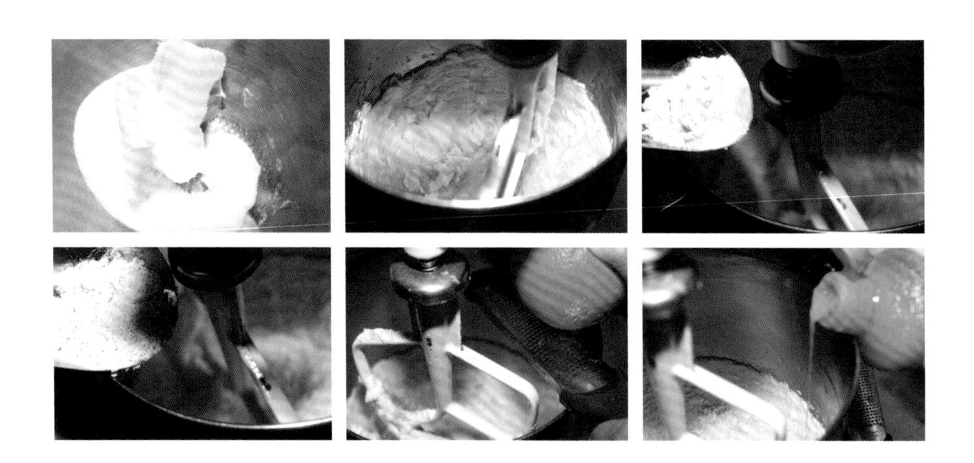

Ingredients

Butter	250 g
Sugar	250 g
Egg	6 pcs
Cake flour	150 g
Almond flour	350 g
Baking powder	5 g
Orange zest	A little
Concentrate orange juice	150 g

Procedures

1 Mix sugar and butter, add eggs one by one to cook.

2 Add cake flour, almond flour, baking powder, orange zest and concentrate orange juice to mix.

3 Put the mixture into the pan and bake for 25 minutes at 170℃.

4 The cooled cake is made into a round shape with a white chocolate.

李　亮 | 面包房厨师长　沈阳碧桂园玛丽蒂姆酒店

Li Liang | Pastry Leader Chef　Shenyang Garden Matitim Hotel

西餐专业委员会会员，2002年从事西饼房工作至今。曾任职于洲际酒店集团、万豪酒店集团、家乐福超市公司、皇朝万鑫酒店、莱星顿酒店、玛丽蒂姆酒店集团。擅长传统西饼、面包制作。

Member of Western-style Food Professional Committee, work in the West bakery since 2002. Once served in the InterContinental Hotels Group, the Marriott group, Carrefour supermarket company, Royal Wanxin Hotel, Lexington Hotel, and Matitim Hotel Group. Skilled in traditional west cake and bread cooking.

黑森林蛋糕 Schwarz Wald Cake

袁晓楠

Yuan Xiaonan

西餐专业委员会会员，从业20年，擅长欧式面包的制作，旁通法式甜品，曾担任多家社会饼店的技术顾问。

Member of Western-style Food Professional Committee, worked for 20 years, skilled in making European bread and French dessert, the consultant of many bakery shops.

材料 牛奶30克，糖16克，蛋黄8克，鱼胶3克，黑巧克力80克，淡奶油120克。

Ingredients

Milk	30 g
Sugar	16 g
Egg yolk	8 g
Fish gelatin	3 g
Black chocolate	80 g
Unsalted butter	120 g

制作方法

1 将30克淡奶油和牛奶煮开，再冲入加有糖的蛋黄中。

2 加热到85℃。

3 加入鱼胶和黑巧克力搅匀。

4 加入打发的淡奶油。

5 倒入模具中，冷冻成型，装盘搭配巧克力酱和酸樱桃即可。

Procedures

1 Boil 30g of unsalted butter and milk, and add sugar into the egg yolk.

2 Heat to 85℃.

3 Stir with fish gelatin and black chocolate.

4 Add the whipped unsalted butter.

5 Put into the mold, freezing, and serve with chocolate sauce and sour cherries.

巧克力慕斯 Chocolate Mousse

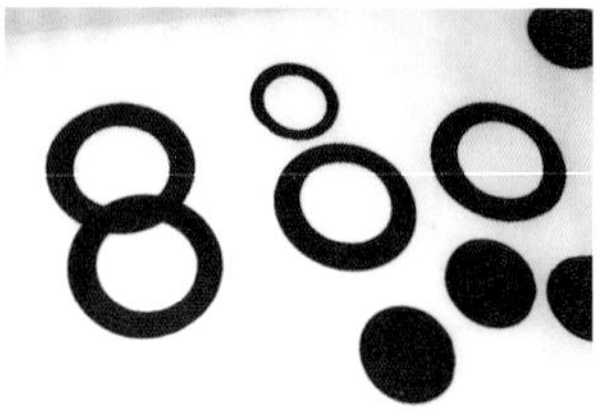

材料 糖230克，牛奶380克，蛋黄300克，巧克力600克，鱼胶40克，淡奶油900克。

制作方法

1 先用融化的巧克力制成巧克力圆片备用。

2 将煮开的牛奶倒入混合的蛋黄和糖中。

3 回火煮到85℃，再加入融化的巧克力，降温到60℃。

4 加入泡好的鱼胶片。

5 降温到40℃，加入打发的淡奶油，倒入裱花袋中冷却。

6 装盘，将做好的巧克力慕斯挤在巧克力圆片中即可。

Ingredients

Sugar	230 g	Chocolate	600 g
Milk	380 g	Fish gelatin	40 g
Egg yolk	300 g	Unsalted butter	900 g

Procedures

1 Make chocolate chips with melted chocolate to set aside.

2 Boil the milk , put into a mixture of egg yolk and sugar.

3 Heat to 85℃, add melted chocolate, and then cool to 60℃.

4 Add soaked fish gelatin.

5 Cool to 40℃, add to foamed unsalted butter and put into the decorating bag to cool.

6 Dish up, squeeze the mousse onto the chocolate chips.

徐　跃 | 饼房厨师长　沈阳凯莱酒店

Xu Yue | Leader Pastry Chef　Shenyang Kellen Plaza Hotel

西餐专业委员会会员，10多年星级酒店工作经验，包房、饼房双修，经验丰富。

Member of Western-style Food Professional Committee, 10 years' experience in star hotels, skilled in both the serving room and pastry, experienced.

核桃派 Pecan Pie

材料 糖1000克，淡奶油350克，核桃仁100克。

制作方法

1 将糖炒成焦糖色，加入淡奶油搅拌均匀。

2 加入核桃仁，降温后放入塔壳中。

3 装盘，搭配焦糖汁和糖花即可。

Ingredients

Sugar	1000 g
Unsalted butter	350 g
Walnut kernel	100 g

Procedures

1. Sauté the sugar into coke sugar, and add the unsalted butter evenly.
2. Add the walnut kernel, and put in the tower shell after cooling.
3. Dish up and serve with the caramel sauce and sugar flowers.

芦光宇 Lu Guangyu | 区域经理、技术顾问、培训导师　美国维益食品公司
Regional Manager, Technical Consultant, Training Supervisor　American VE Food Companies

1994年从事烘焙行业，先后在香港新世界酒店管理集团和香格里拉酒店管理集团成员的多家星级酒店担任西点主厨职务。2006年取得全国高级烘焙技师专业资格证书。

Be engaged in baking in 1994, served as pastry chef in a number of star hotels such a as Hong Kong New World Hotel Management Group and Shangri-La Hotel Management Group. Get the certificate of national professional bakery in 2006.

胡萝卜蛋糕 Carrot Cake

材料 蛋黄、蛋清各5个，糖260克，杏仁粉225克，蛋糕粉130克，泡打粉2克，胡萝卜末375克。

制作方法

1 分别将蛋黄、130克糖和蛋清、130克糖打发，再混合均匀。

2 加入杏仁粉、蛋糕粉、泡打粉搅匀。

3 加入胡萝卜末搅匀。

4 用180℃烤制25分钟即可。

Ingredients

Egg yolk	5 pcs
Egg white	5 pcs
Sugar	260 g
Almond powder	225 g
Cake flour	130 g
Baking powder	2 g
Carrot pieces	375 g

Procedures

1 Respectively stir evenly to make foam of egg yolk, 130g sugar and egg white, 130g sugar sent.

2 Mix almond powder, cake flour, baking powder.

3 Add carrot pieces and stir.

4 Bake for 25 minutes at 180℃.

谭龙巍 | 私营业主

Tan Longwei | Private Owner

西餐专业委员会会员，拥有在多家五星级酒店任职的经历。擅长法式甜品的制作，尤其对翻糖蛋糕有独到的见解。

Member of Western-style Food Professional Committee, once worked in many five-star hotels. Skilled in the production of French dessert, with in particular the unique view of the fondant cake.

双色咖啡蛋糕 Double Color Cake

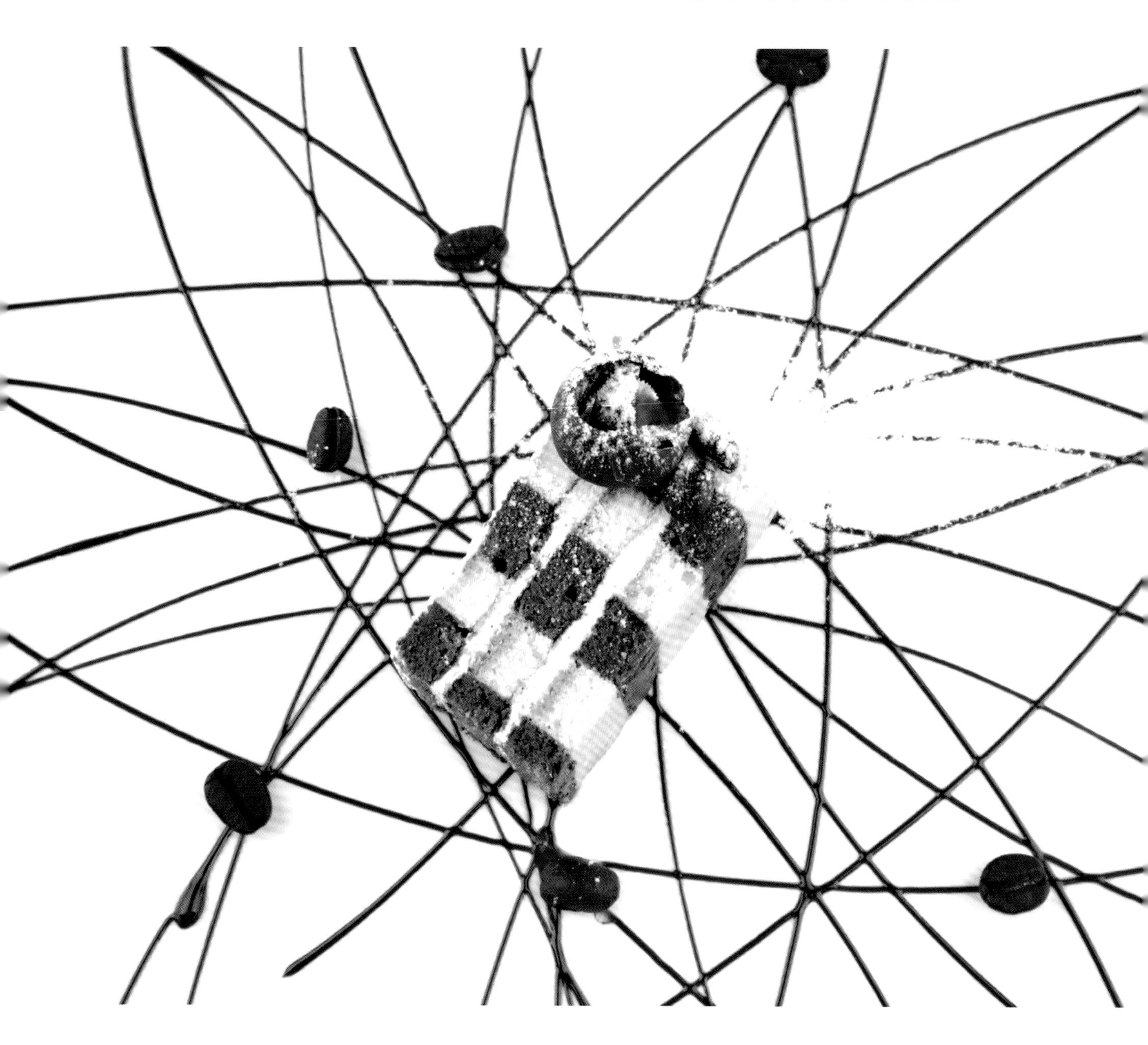

阮　详 | 技术主管　铁岭爱优美烘焙坊

Ruan Xiang | Technique Directive　Ai You Mei Bakery in Tieling

西餐专业委员会会员，荣获第十六届全国焙烤职业技能竞赛《维益杯》辽、吉赛区选拔赛金奖，第十六届全国焙烤职业技能竞赛《维益杯》全国决赛铜奖。

Member of Western-style Food Professional Committee, win a gold award in the Weiyi Cup of the 16th national baking skill contest in Liaoning and Jilin region, a bronze award in the Weiyi Cup of the national finals of the 16th national baking skill contest.

材料　黄油、糖、鸡蛋、蛋糕粉各100克，可可粉30克。

制作方法

1 将黄油、糖打发，加入鸡蛋打匀，再加入蛋糕粉搅拌均匀。
2 将一半的料留出备用。
3 将另一半料加入可可粉拌匀。
4 将两种面糊分别装入裱花袋，再交替挤成黑白相间的蛋糕坯。
5 用170℃烤制20分钟，取出冷却。
6 装盘，搭配咖啡豆和巧克力酱即可。

Ingredients

Butter	100 g
Sugar	100 g
Egg	100 g
Cake flour	100 g
Cocoa powder	30 g

Procedures

1 Stir the butter and sugar, and add egg, add cake flour.
2 Set aside half of the material.
3 Add the cocoa powder into the other material and mix.
4 Put two types of batter into a pastry bag and alternatively squeeze into same color of white and black.
5 Bake for 20 minutes at 170℃ and take out for cooling.
6 Dish up and match with coffee beans and chocolate cream.

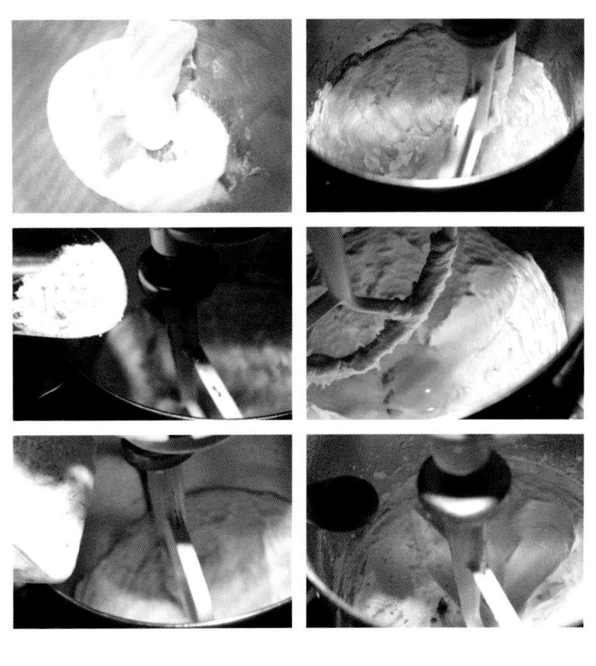

手工冰淇淋 Handy Ice-cream

材料 蛋黄13个，糖180克，蜂蜜200克，芒果果蓉400克，淡奶油750克。

制作方法

1. 将糖、少许水放入锅中煮开，再冲入到打发的蛋黄中。
2. 加入蜂蜜、芒果果蓉。
3. 放入打发的淡奶油混合均匀。
4. 冷冻后食用即可。

Ingredients

Egg yolk	13 pcs
Sugar	180 g
Pure honey	200 g
Mango fruit puree	400 g
Unsalted butter	750 g

Procedures

1. Put the sugar and little water into the pot, pour into the foamed egg yolk.
2. Add pure honey and mango fruit puree.
3. Put into the foamed unsalted butter and mix.
4. Eat after freezing.

闫　寒 | 技术主管　铁岭市聚利来食品工厂

Yan Han | Technique Director　Ju Li Lai Food Factory in Tieling

西餐专业委员会会员，出生于1986年，2002进入烘焙行业，擅长法式面包、欧式面包和日式软面包。

Member of Western-style Food Professional Committee, born in 1986, start baking since 2002, skilled in French, European bread and Japanese soft bread.

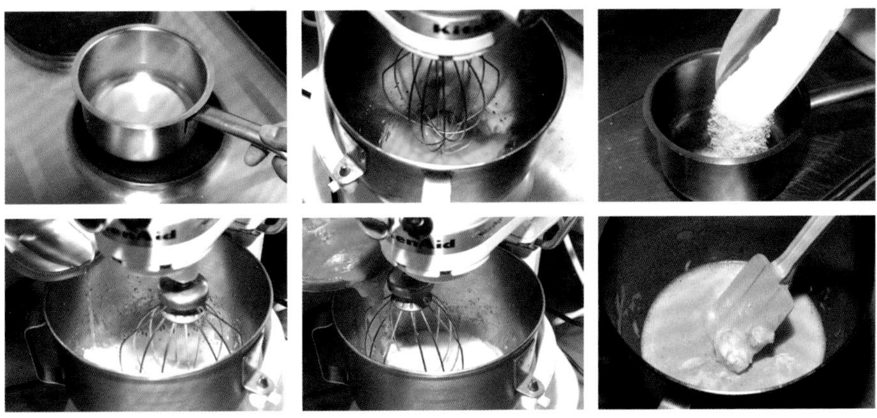

费南雪蛋糕 Financies Cake

赵　亮
Zhao Liang

高级西点技师，现在广州W酒店与法国面包师共事。
Senior Western Pasty Cook and Work together with French bakers in Guangzhou W Hotel.

材料　蛋清150克，糖220克，杏仁粉110克，蛋糕粉65克，黄油（化）190克。

制作方法

1 容器中依次放入蛋清、糖、杏仁粉和蛋糕粉混合均匀。
2 加入黄油混合均匀。
3 将面糊倒入模具中，放入烤箱中，用170℃烤20分钟。
4 取出冷却，用新鲜水果、干果仁和巧克力碎片装饰搭配即可。

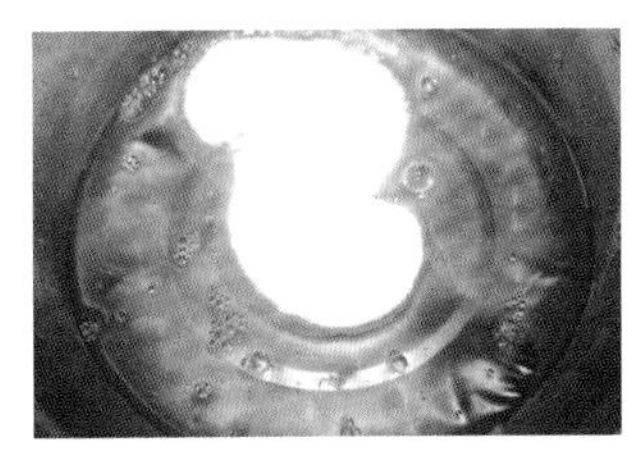

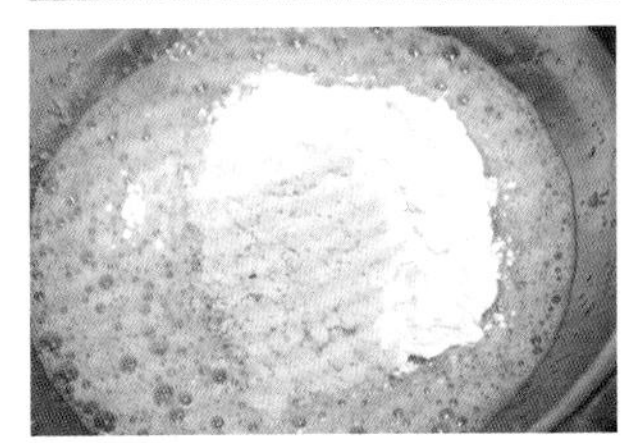

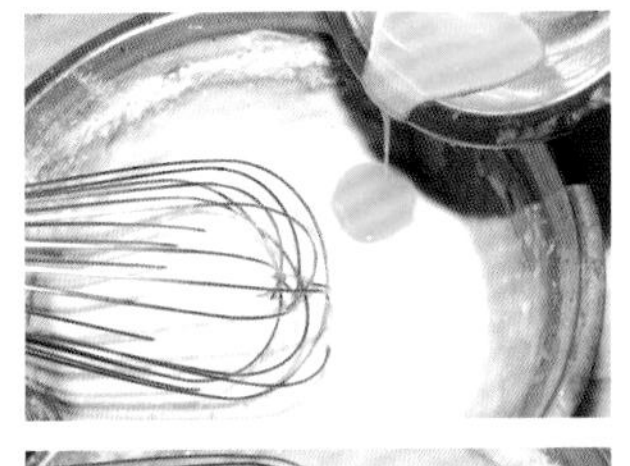
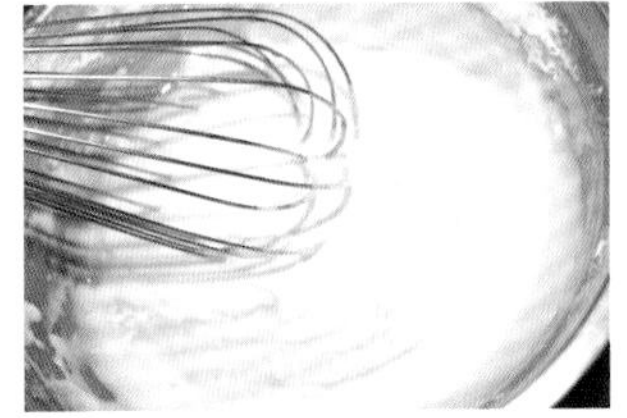

Ingredients

Egg white	150 g	Cake flour	65 g
Sugar	220 g	Butter(Melted)	190 g
Almond powder	110 g		

Procedures

1 Put egg white, sugar, almond powder and cake power in turn and mix them to a smooth paste.
2 Add the butter and stir them.
3 Put the paste in a mould, and put it into the roaster and roast it at 170℃ for 20 mins.
4 Take out the cake and cool it down. Then decorate it with some fresh fruit, dry nutlets and some chocolate chips.

红糖坚果蛋糕 Brown Sugar Nut Cake

材料 黄油100克，糖75克，鸡蛋90克，蛋糕粉、牛奶、枸杞各10克，泡打粉2克，杏仁粉20克，红豆50克。

Ingredients

Butter	100 g
Sugar	75 g
Egg	90 g
Cake flour	10 g
Milk	10 g
Medlar	10 g
Baking powder	2 g
Almond powder	20 g
Jumby bean	50 g

制作方法

1. 混合黄油和糖，打发。
2. 加入鸡蛋搅拌均匀。
3. 加入蛋糕粉、泡打粉、杏仁粉、牛奶、红豆和枸杞混合均匀。
4. 将蛋糕糊倒入烤盘中，在175℃的烤箱中烤制25分钟。
5. 取出蛋糕冷却，切成长方形，装盘，搭配红糖和少许坚果即可。

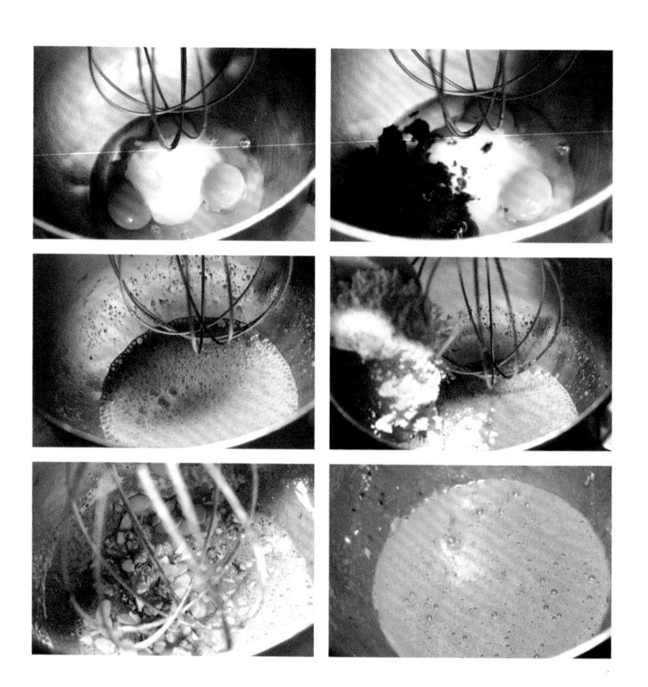

Procedures

1. Mix the butter and the sugar to fermentation.
2. Add the eggs and mix them together.
3. Add cake power, baking power, almond power, milk, jumby bean and melar and mix them to a smooth paste.
4. Put the cake batter on ovenware and roast them at 175℃ for 20 mins.
5. Take out the cake and cool it down. Cut it into a rectangle, put it on a tray and accompany it with some brown sugar and a few nuts.

刘　双 | 逸林饼房高级主管　沈阳希尔顿

Liu Shuang | Senior Executive of Yi Lin Bakery　Shenyang Hilton Hotel

曾任职于玛丽蒂姆酒店集团、洲际酒店集团、希尔顿集团等多家集团酒店。

Once worked in Matitim Hotel Croup, InterContinental Hotels Group (IHG), Hilton Hotel Croup, and etc..

Westem Food Classroom II

西餐教室II